The Natures of Science

The Natures of Science

Neville McMorris

Rutherford • Madison • Teaneck
Fairleigh Dickinson University Press
London and Toronto: Associated University Presses

Associated University Presses
440 Forsgate Drive
Cranbury, NJ 08512

Associated University Presses
25 Sicilian Avenue
London WC1A 2QH, England

Associated University Presses
P.O. Box 488, Port Credit
Mississauga, Ontario
Canada L5G 4M2

The paper used in this publication meets the requirements of the American National Standard for Permanence of Paper for Printed Library Materials Z39.48-1984.

Library of Congress Cataloging-in-Publication Data

McMorris, Neville, 1934–
   The natures of science.

   Bibliography: p.
   Includes index.
   1. Science—Philosophy.   2. Science—Methodology.
3. Science—History.   I. Title.
Q175.3.M393   1989   501   87-46376
ISBN 0-8386-3321-8 (alk. paper)

Printed in the United States of America

**For Mary, Mother, and Kevin,
Julian, and Nicolas**

But the images of men's wits and knowledges remain in books, exempted from the wrong of time, and capable of perpetual renovation. Neither are they fitly to be called images, because they generate still, and cast their seeds in the minds of others, provoking and causing infinite actions and opinions in succeeding ages.

—Francis Bacon, *The Advancement of Learning*

Contents

Preface

> . . . No cognitive area *rests* its case on any single source of knowledge. . . . And it is certainly wrong to say that . . . science rests its case on experience or experimental evidence. Science does *not rest* its case on any one source of knowledge, but experience and experimental evidence, and clear and distinct ideas, aesthetic considerations, and analogy are all legitimate sources of knowledge.
>
> —Y. Elkana, "Science as a Cultural System"

This book, which is an extended essay more than a tome of research, arose nevertheless out of continuous reflection over a period of many years during which time some of the material now presented was published in historical and philosophical journals.

It is more of an essay because, despite both the many diverse areas treated in this work, and the close arguments used in support of some theses, an attempt has been made to present it with an underlying theme and, therefore, as a continuous piece both structurally and conceptually. In striving to achieve these aims I have frequently adopted a narrative and allusive form, aided by numerous references, back and forth throughout the book, to the organic links between the natures of science being considered. The aims of this work have dictated also that there be much rewriting of the already published papers; and this exercise has, I hope, reduced the annoying breaks and repetitions normally found accompanying collections of published papers. Thus, where this work is not newly written, it has been largely rewritten.

Chapters 9 and 10 are entirely new, and Chapters 1 and 4 are vastly changed in language, in organization and in opinions and arguments, from the original. Much new material has also been added. Chapters 3 and 5 have been extensively reorganised also, and the rewriting has been substantial. The changes effected in Chapters 7 and 8 have been fewer, mostly of an organizational and editorial nature, although some new material has again been added; the changes in Chapters 2, especially, and 6 have been even fewer. Despite the rewriting, many of the chapters now presented have retained their original titles.

The idea of a book arose from the realization that many of my written and projected papers were falling without premeditation, into what appeared to me to be natural divisions, yet all related through an underlying seam that I have labeled "The Natures of Science." There is a large scale as well as a finer division among these chapters. The larger may be listed as (I) Philosophical, (II) Aesthetic, (III) Cultural, (IV) Methodological, and (V) Scientific. Within each of these parts, spanning two chapters, is a division into "general" and "particular." Thus the first chapter in each section discusses in relatively general terms the topic of that section, whereas the second chapter may be seen to exemplify the related scientific nature through particular theories, systems, or episodes. Perhaps there is an exception to this in Part III, in the essay on Eliot and Einstein.

Even though particular examples exist of the methodological impact of science within Science (Galileo), and outside of Science (History), it seemed to me more interesting to keep this essay, which employs scientific concepts rather than scientific method as the stimulant, and poetry as the context.

"Method" in Galileo and History appear elsewhere in the book and the introduction of Poetry makes the linkages of Science that much richer.

The choice for the last chapter of the book was made with the intention of giving shape to the whole by reinforcing the continuity and integrity of the view of science scattered through the previous chapters. At the end as at the beginning, in Chapter 10 as well as in Chapter 2, we have the concept of duality and its relation to the physics of light. We might say that it was treated at the beginning philosophically and at the end scientifically. This description will attract some people who would hold that it reveals the natural and historical development of science out of philosophy, but although the historical development has been generally proved, the naturalness has not been so demonstrated. Be that as it may, I am prepared to let that chapter stand as an example of the perennial relatedness and relevance not only of philosophical ideas to science but, by extension, of aesthetic, cultural, and methodological ideas as well. After all, this is the point of the essay.

Even so we have to begin by dealing with that apparent monolith that we call Science.

"Science" and "scientific" are terms promiscuously applied to a wide range of activities and systems so that "scientific" is a state that we all claim for ourselves, and "science" is a label that we all aspire to attach to our disciplines even while controversy rages over the validity of these aspirations. Thus the athlete and the critic will regard themselves as being scientific in their approach to training and to texts, respectively, whereas the sociologists whose discipline is found under the rubric of "social science" publicly agonize whether their subject is a science. Some physicists, on the other hand, know that their subject is a science; others, more arrogantly, that it is Science. That physics is a science is universally accepted.

An interesting sociological phenomenon associated with these confident claims and disciplinary uncertainties may readily be explained within our present cultural context as well as historically. There is clearly widespread consensus about what *is* science but very little about what is not. It is equally clear that the image of science is alluring as well as bright; indeed physics has established itself in contemporary consciousness as intellectually challenging; as providing an explanatory matrix for many of our ancient perplexities; and as being dramatically and socially beneficial. It is doubtful if any other discipline has achieved such success over such a wide area of human concerns. For others to aspire to have their discipline acquire the label of sceintific is therefore perfectly natural.

The fact that most scholars might wish to have their subject or area of activity regarded as a science arises also from an interesting historical perspective on the word, and the concept of, "science." This perspective reveals as well philosophical and methodological aspects of science. If we concentrate on the words "science," *"scientia"* as the medievalists would write it, or *"episteme"* as Aristotle did write it, we hit upon a continuous thread that leads us to see the importance that "science" has enjoyed throughout more than two millennia. The word meant for Aristotle and his disciples knowledge but it also meant a special kind of knowledge, the highest to which the human mind could aspire— knowledge that investigated the causes of things. We therefore have always aimed at scientific knowledge, and this fact gives historical depth to the sociological phenomenon just mentioned. Furthermore, *"episteme"* and *"scientia"* were concerned with the acquisition of demonstrative knowledge, that is the most certain knowledge, so that if the twentieth century aspirations might occasionally appear neurotic this is only because one particular area of knowledge seems to have acquired techniques that have been more successful at achieving this long-standing and Faustian ambition—which is one of the main theses of this essay.

Science, therefore, through the demonstrative syllogism of Aristotle bore a methodological imprint that it prominently displayed for two thousand years until Francis Bacon fully articulated that long-simmering experimental methodology that was influential in displacing Aristotle's. At the same time the methodological shift could be described as a change from the Aristotelian concern with "why" questions to Galileo's concern with "how" questions, or as a shift in emphasis from "thinking" to "doing."

We are led by these changes into others that were of equal importance, namely the philosophical and the cultural. Aristotle's "science" was embedded in philosophy to the extent that his philosophy arose from its fundamental interest in the causes of things; and the concern of his science with "why" questions also reveals such a philosophical orientation. The methodological moves made by Galileo and by Francis Bacon gained some of their significance because, it was thought, they largely helped to achieve disengagement of science from philosophy. We add, however, that the disengagement was partial because even Bacon's science had a philosophical framework. But whereas Aristotle's science was identified with philosophy, and not only linguistically, Bacon's science was merely nourished by it. One of the major arguments of this essay is that no permanent or total disengagement is at all possible.

The cultural track on to which science was led from the seventeenth century to the twentieth may be described most aptly as methodological, to the extent that science's most profound impact upon the widest spectrum of society has been through its methods or, to put it more tellingly, through its *successful* methods that encompass objectivity, quantification, and generalization. The cultural component of science, educed in this way, takes us back to the sociological phenomenon, so described, in our contemporary society. If the traits mentioned previously are those inculcated in us by science, then again there is no surprise that we would all wish to be scientists.

Still within the realm of the cultural, although the impact might be less wide and profound, is the realization that nature provides through science evidence of its essential order and simplicity. It is true that humanity has tried to impose order and simplicity on nature, but the wonderful thing is that nature has responded, thus revealing the extent to which science is supported by aesthetic elements.

This preface has teased out the nature of science in such a way as to introduce the idea, in a natural way, that science has several natures that may be described according to the headings of the parts I have listed. It has been done in this way to encourage the view that I hold about the nonmonolithic nature of science—that, indeed, there is more to science than its being scientific. The arguments for the case are presented in the following chapters.

These arguments should be of interest to a very wide range of people. Academically it is aimed at those specialist students in the history and philosophy of science as well as at those liberal arts students interested in science. Outside of the university context the book should find an audience among those of broad education and interests; those who are interested not only in philosophy or literature or education and culture, or science or history, but in all intellectual sciences. The essay should break down any continuing divisions, either perceived or real, among these areas of knowledge, and if the title of the essay is taken seriously, and the message is communicated, then all areas will indeed be seen not merely as aspects of knowledge but as an integrated body of knowledge. It is hoped, finally, that the essay will fall into the hands of scientists also who will assuredly view their avocation or their profession rather differently when seen in the light of these many and closely related natures of science.

Acknowledgments

It is easy and pleasant to record all acknowledgments in connection with the preparation of this book. They all go to Mrs. Ditta McDonald, the departmental secretary, who has not only typed and retyped many versions of the manuscript, but has had to work from the most unpromising scribble. She has also succeeded in pointing out occasional errors or infelicities in the manuscript, and in questioning me on the meaning of some sentences, which forced me to refashion them. Thus has she exhibited both patience and imagination, as well as skill. I am fully and genuinely grateful for her help.

The acknowledgment for critical moral support goes entirely to my wife who suggested an appropriate attitude to adopt in the writing of this book, and thus encouraged me to continue with a venture when it had stalled.

I acknowledge the kind permission of the following periodicals to use material that has appeared in their publications. Chapters 1, 7, and 8 have appeared in some form in *Physis;* Chapters 2 and 4 (especially) have appeared in very different form in *Scientia;* and the substance of Chapters 3, 5, and 6 has appeared in *Main Currents in Modern Thought.*

I acknowledge also the permission of the following publishers to quote from their editions of the poems of T. S. Eliot:

Faber and Faber Ltd.: *Poems Written in Early Youth; Collected Poems 1909–1962*

Farrar, Strauss and Giroux: *Poems Written in Early Youth*

Harcourt Brace Jovanovich: *Collected Poems 1909–1962; The Four Quartets*

The Natures of Science

PART I
The Philosophical Nature of Science

1
Science as *Scientia*

... since there is no science which is not some part of philosophy, we should see therefore, what philosophy is.

—Domingo Gundisalvo, "Classification of the Sciences"

The use of the word *scientia,* as if it differed from knowledge is a modern barbarism. . . .

—John Ruskin, *Works*

Introduction

The philosophical nature of science can be investigated through philosophical analysis, and this has been done often and well. In such exercises it is tacitly accepted both that science and philosophy are autonomous disciplines, and that science, through its concepts and theories, has philosophical implications. This is no special mark of science, for other disciplines have such implications. There is, however, another approach through which science may be seen to be special, to be essentially philosophy. It will readily be accepted that this approach must be different, for far from assuming that science and philosophy are autonomous disciplines, it accepts them to be almost indistinguishable. This approach involves historical and linguistic analysis, and the pertinence of such an analysis will be appreciated when it is realized that an important facet of science is embedded within the etymology of the very word "science," which has philosophical undertones.

Of course science, as it is practiced in the twentieth century, has only the denotation but not the connotation of *scientia.* To this extent, therefore, it might seem unrealistic, indeed anachronistic, to talk about the philosophical nature of science. However, even if the origins of a word, an idea, or a system have been obscured and distorted by time and practice, at the very least some of its primitive meaning remains of perennial interest; and we shall be concerned with recovering these origins. Such recovery of the fundamental meaning of science, through historical analysis, should enlighten the educated community about the wider and deeper significance and potentialities of science; and at the same time it should stimulate activity that would embody this significance. Our concern, however, plumbs profounder depths than the merely historical, for we shall argue that the philosophical nature of science has only been obscured.

It is to be emphasized that although *scientia,* associated with the Greek equivalent *episteme,* might have the import of universal knowledge that philosophy takes for its province, science has not been viewed generally in this way for three centuries. It is indeed argued, as a substantive issue in this chapter, that what characterizes the perceptions of the difference between pre- and post-seventeenth-century science is that before the seventeenth-century revolution, science is seen to have been *scientia;* that is, it was only one part of the overarching, philosophical concern with the world. After the revolution it became an autonomous activity; it became science.

Philosophical, methodological, and linguistic arguments support this characterization of the earlier science, which stretches from the time of Aristotle to the thirteenth century when hints of a special, independent view of science first emerged through the work of Roger Bacon. I believe that the twelfth-century scientific revolution first stimulated this new and distinct view of science as glimpsed in Roger Bacon.

But it might be argued that although the perspective of those subjects we now call science might have been affected by Bacon's writings, the change in the use of the word *scientia,* as glimpsed in Bacon, was not sustained. To this extent, therefore, science remained *scientia* up to the seventeenth century, after which time the perspective of science is regarded as having definitely changed.

That is the historical truth. In a fundamental way, however, science has not forgotten its origins that reside firmly in the *episteme* of Aristotle, for it still performs the major philosophical task of allowing us access to the greatest degree of epistemological clarity and certainty.

Science and *Scientia*

This chapter therefore proceeds to explore the prehistory of the word "science" and it can hardly do better than to refer, at the start, to the fascinating study by Ross, who argues for the sociological need in the nineteenth century for the invention of the word "scientist." He documents his historical study with several excerpts from letters and other sources of personal reminiscences whose authors expressed reservations about the introduction of this linguistic hybrid. Ross further supplements his study with the help of detailed philological comments.

The main point of departure for our purpose, however, is Ross's consideration of the evolution of the word "science." He supports the frequent assertion that *"Science* was synonomous with *knowledge,"* that the "sciences . . . were specialized branches of philosophy." Ross further identifies the period from 1620 to 1830 as one during which there occurred "a . . . change in the significance accorded to the word science." Indeed he thinks that Carlyle, in 1829, correctly interpreted a trend in which "the word *science* in common speech came to have the dominant meaning of 'natural and physical science.' "[1] From this two issues immediately arise, namely the accuracy of identifying "science" with mere "knowledge" and the timing of the historical evolution of the word "science" itself.

As to the first issue the fact behind the assertion seems simple, for as Ross himself reminds us "Science entered the English language in the Middle Ages as a French importation synonomous with knowledge." The French word itself, *science,* derived as it was from the Latin *scientia,* whose fundamental meaning was knowledge, shows some justification for the assertion. Indeed it is known that "science" when it was *first* imported into the English language had "knowledge" as one of its commonest synonyms, and we shall give appropriate examples of this usage. But *"scientia"* from which it was derived meant more than mere knowledge; thus, although it might be true to say that "science" once meant knowledge, this claim is not equally true of *"scientia."* We shall elaborate on this in the next section.

The second issue will be taken up more fully in Chapter 9; nevertheless, it is relevant for us that despite the nineteenth-century changes adumbrated by Ross, there were those persons, even in that century, who clung to the original and narrow meaning of "science" and, indeed, of *"scientia."* Ruskin was one of these, an intellectual anachronism. For him having casually ignored two centuries of evolution of the word "science," there was nothing peculiarly scientific about chemistry or physics. When we link this view with another of his that "The use of the word *"scientia",* as if it differed from knowledge, is a modern barbarism,"[2] we arrive at Ruskin's (narrow) conception of Science as *Scientia,* which has nevertheless served as the title of this chapter. We therefore superficially agree with Ruskin, but differ profoundly with him in believing that "science" arose from a word that connoted more than mere knowledge.

Methodology

A summary account of the scope and approaches of this chapter finds us firstly examining the use of *scientia* in Aristotle, which carried very often, through the qualifier "demonstrative," the import not so much of an activity as of a system with his own philosophical coloring. As noted elsewhere in this chapter, an active recognition of such philosophical moorings will lead to the development within science of richer perspectives of its present methodology.

We argue, by drawing on the work of Stock, Stiefel, Ross, and others,[3] that after a minor revolution in the thirteenth century the first signs of growth of science were detected. It readily follows (from certain historiographical vantage points) that a second, transitional period would arise before the modern usage of "science" became firmly established within a third period. Ross accepts the first period and cursorily examines the others. These latter are of great interest and will engage our attention in later chapters. In this chapter we shall be concerned with the first period, but where Ross incidentally accepts it, we shall explore the validity of this acceptance, for it is from the establishment that science was *indeed* philosophy that we are hoping to maintain that one of its *essential* natures is philosophical.

A brief examination follows these considerations in which comments by writers of the medieval period are discussed; and particular attention is paid to those decades during which Aristotle's works were recovered and a deep and pervasive concern with science raised itself again in the West. A consonance of views between these writers and Aristotle will demonstrate the implied and unsurprising continuity of tradition and will also provide validation for our omission of so many centuries of history: we shall find that indeed science was still *scientia.* Despite the glimpse of changes that an examination of Roger Bacon's philosophy and language allows us, we conclude nevertheless that the changes were neither pervasive nor sustained.[4]

Before we launch into the establishment of our main thesis,

however, we must clarify what we mean by the word "science" and by the phrase "activity of science." To the extent that science is about the knowledge and understanding of nature, it has existed for millennia. The ways in which it has been executed have been variously manifested from Aristotelian rationalism through the methodological changes of mathematization and experimentation. Encompassed and fertilized by these changes have been physical and biological topics: painting and politics as well as rhetoric and religion have been excluded. Thus even in ancient and medieval times, physics and biology were science supported by a different methodology, and they were indiscriminately referred to as natural *philosophy,* which gives a clue to the status of science, the activity. At the same time the word *"scientia"* encompassed a wider spectrum so that music and morals as well as poetry and physics were included. To *this* extent only, as we remarked above, *scientia* was mere knowledge and, by implication, so was science. But in Aristotle, *episteme* (and in others, *scientia*) stood in place of *scientia demonstrativa,* which we shall indeed argue; and for the larger case being made in connection with the philosophical entanglements of early science, this qualification is necessary. "Knowledge" as the merely passive, dead meaning of *scientia* can bear no philosophical fruits. Thus in our title, "Science as *Scientia,*" the word "Demonstrativa" is to be understood, as it often was in Aristotle.

Aristotelian Science

PHILOSOPHICAL BASIS

Aristotle's scientific work[5] must be taken seriously by anyone interested in the history and philosophy of science because it was the first work to encompass and treat in a profound way the whole range of scientific topics and subjects that stimulated philosophers in his time and because there is a clear methodological commitment informing Aristotle's treatment. Furthermore, although we are now aware of the many ways in which he was wrongheaded we cannot discount the enormous influence that his works had in many seats of learning over many centuries.

We start with the philosophical basis of Aristotle's science for as Aristotle had suggested, in part, and as Domingo Gundisalvo was to say, most appositely as late as c. 1140 "since there is no science which is not some part of philosophy, we should see therefore, what philosophy is." So science continued to be seen as integrated with philosophy over a period of more than fifteen centuries; it had no other basis.

Aristotle opened his *Metaphysics* with the observation that "All men by nature desire to know."[6] As a philosopher—that is, a lover of wisdom or knowledge—his desire was even greater. He was concerned, as we have said, with all knowledge, which he divided into theoretical, practical, and productive—an example of each being mathematics, politics, and sculpture. Everything that could be known was included in this classification. Within this comprehensive framework Aristotle saw the main thrust of his philosophy as gaining *certain* knowledge: the thrust was epistemological. It was this requirement of certainty that separated subjects and made some more and others less "scientific"; but they were all "science."

The method that he used to gain this certain knowledge was the syllogism, his logical tool. Indeed, he regarded a philosopher as someone who studied the nature of matter and the principles of syllogism.[7] The syllogistic method with its underlying demonstration was important for his purposes because this method was employed by geometry, the epitome of certain knowledge. Geometrical demonstration followed from evidently true premises to *certain* conclusions and was capped by the triumphant *"Quod erat demonstrandum."*

All such *demonstrative* knowledge was for Aristotle true or what he might have called "scientific" (or more certain) knowledge. To quote him: "We suppose ourselves to possess unqualified scientific knowledge of a thing when we think that we know the cause on which the fact depends . . . what I now assert is that at all events we do know by demonstration. By demonstration I mean a syllogism productive of scientific knowledge. . . . Thus the premises of demonstrated knowledge must be true, primary, immediate, better known than and prior to the conclusion. . . ."[8]

At the same time Aristotle dismisses other kinds of knowledge, like the dialectical, because they are inadequate for the production of that certainty which he saw as the hallmark of scientific knowledge. This is so because "If a proposition is dialectical, it assumes either part indifferently; if it is demonstrative, it lays down one part to the definite exclusion of the other because that part is true."[9] For him scientific knowledge could not arise from any impartial balancing of alternatives or opposites; it arose from a clear-sighted view of truth via premise to demonstrated conclusion. Even so he required that the premise be necessary as well. "For though you may reason from true premises without demonstrating, yet if your premises are necessary you will assuredly demonstrate—in such necessity you have at once a distinctive character of demonstration."[10]

As noted, Aristotle divided the kingdom of knowledge into three parts: theoretical, practical, and productive.[11] It so happened that, according to his classification, it was theoretical knowledge that could achieve greatest certainty and therefore the greatest epistemological status. It is this nature of theoretical knowledge that Aristotle is talking about in his *Metaphysics* when he asserts that theoretical rather than productive kinds of knowledge are more of the nature of Wisdom.[12] These theoretical sciences were Metaphysics, Mathematics, and Natural Science. Thus, Natural Science (or Physics) was simply an area of knowledge, like others, that was subjected to the overriding philosophical demand of syllogistic demonstration. For Aristotle, therefore, Mathematics, Physics, and Metaphysics could be more certainly demonstrated than other subjects and this feature qualified them as more "scientific." In this context physics, a theoretical science, lay somewhere in the continuum between the demonstrated certainty of geometry and the gross uncertainty of politics (a practical science), being much nearer to the former.

Therefore, as Aristotle saw it, what we call science was not an activity different from philosophy. That this is so is further indicated when Aristotle in his *Physics* states that "knowledge is the object of our enquiry and men do not think they know a thing till they have grasped the 'why' of it."[13] This concern with the "why" was central to Aristotle's philosophy for "a grasp of a reasoned conclusion [i.e., an answer to the question why?] is the primary condition of knowledge." He sharply dismissed the notion that knowledge was merely experience, for "men of experience know the thing is so, but do not know why, while the others [artists or scientists] know the 'why' and the cause."[14] Reference has often been made to this concern of Aristotle's. We simply cite one of them, by Edelstein who remarks that "Physics . . . in antiquity remained closely connected with philosophy, and was predominantly concerned with the philosophical category of the 'why,' rather than with the scientific category of the 'how.' "[15]

Science was not a separate activity but was part of the epistemological concern of philosophers. It was indeed *episteme.*

LINGUISTIC BASIS

Episteme apodeiktike

Only an inexhaustive search through Aristotle's original texts in Greek is necessary to reveal that the word for knowledge is ἐπιστήμη. When we translate Aristotle as saying that "knowledge . . . of the fact and of the reasoned fact, as contrasted with knowledge of the former without the latter, is more accurate and prior," the word translated as "knowledge" is *episteme,* which makes sense in the context of the discussion of epistemology found in his *Posterior Analytics.*[16] We can be satisfied therefore (and it is the case) that *episteme* is what Aristotle is also talking about when he says that knowledge is the object of our enquiry, in the quotation from his *Physics* cited previously. Physics is clearly the activity science, and its inclusion in this concern about an obviously philosophical matter certifies again the fundamental way in which science was integrated with philosophy. However, *episteme, tout court,* does mean knowledge.

As far as the activity of science is concerned we recall that Aristotle dealt with subjects that we would call science. However, when he wished to refer to such subjects he often employed either the term natural philosophy or apodictic (i.e., demonstrative) knowledge: ἐπιστήμη ἀποδειχτική. "Natural philosophy" properly locates the subjects, and "demonstrative knowledge" does it precisely; and this precision is reinforced by Aristotle's disclaimer that "not all knowledge is demonstrative; the knowledge of immediate premises is not by demonstration."[17]

The implication of this qualification is that we must be alert to the fact that although it is argued that "science," *scientia,* and *episteme* meant knowledge, (natural) science was knowledge of things *and* their causes. It is necessary to qualify the occasionally loose statement that science and *"scientia"* in antiquity meant knowledge generally. In fact, science or scientific knowledge was not so general; it was qualified, demonstrative knowledge.

However, *episteme* was used without qualification to mean demonstrative knowledge or, as we would translate Aristotle, "science" or "scientific knowledge," for example:[18] "A science is one if it is concerned with a single genus" and "Scientific knowledge cannot be acquired by sense perception."[19] *Both* "science" and "scientific knowledge" are renderings of *episteme,* which was quite often used in place of the fuller and more precise expression, *apodeiktike episteme.* Thus although *episteme,* from the generality of "knowledge" to the less general "demonstrative knowledge," acquires some specificity of its own, it yet does so within a context that provides the linguistic reference for the philosophical position unfolded in the previous section.

Scientia demonstrativa

It is still left for us, however, to establish a similar usage in Latin through which language the medieval and Renaissance philosophers gained access to Aristotle's science, a word that we have indicated originated from *scientia.* We wish to satisfy ourselves, therefore, that the corresponding usage in Latin was *scientia demonstativa.*

Certainly Boethius,[20] in the sixth century A.D. talks freely about "divine knowledge," "this faculty of knowledge," and of "foreknowledge," all expressions of knowledge being firmly based on *"scientia."* Furthermore the numerous discussions of *scientia* together with *sapientia* (and *prudentia* as well), which are such general virtues, suggest strongly the general connotations that *"scientia"* had for these thinkers.

However, we go immediately to the particular Latin usage in respect of science—that is, what particular qualification was applied to *scientia* in discussions of science—and we turn to Gilbertus Porreta, John of Salisbury, and Gundisalvo for the evidence. Through their works we can savor what was the view of science just about the time when Aristotle's works were being rediscovered and assimilated.

Firstly, Gilbertus is thought to have been vaguely aware of the need for induction from experience through the Aristotelian means of *demonstration.* His pupil, John of Salisbury, tacitly accepted Aristotle's view that not all knowledge was demonstrative. For example, "one must . . . use every effort to fortify . . . his proofs, lest there seem to be, . . . some gap, which would jeopardize . . . demonstrative science [scientiam demonstrativam]. By no means all science, but only that which is based on truths that are primary and immediate, is demonstration . . . although every real demonstration consists in a syllogism. It is the inherent nature of science to strive for demonstration."[21] Although John of Salisbury is commenting on Aristotle, he is at the same time inserting his own views,[22] and we note that some *scientiae* are made special by being demonstrative.

Furthermore, Gundisalvo reveals a similar understanding of

scientia in his *De Divisione Philosophiae.* For him, also, physics is a science, considering things unabstracted and with motion. "This science provides knowledge of natural bodies. . . . The instrument of this art is the dialectical syllogism. . . . The practitioner is the natural philosopher who, proceeding rationally from the causes of things to effects and from effects to causes, seeks out principles."[23] So, again, physics is qualified *scientia.*

His view of physics or natural science becomes clearer when it is recognized that he regards logic as a necessary tool in the study of sciences, where its purpose, on the practical side, includes reasoning that he divided into dialectic and demonstration. For him, too, demonstration proceeded from self-evident propositions. It is therefore the (dialectical) syllogism that produces scientific knowledge by providing through logic that demonstration which Aristotle himself assigned to the highest form of knowledge. For Gundisalvo, who had not even been fully acquainted with the recovered works of Aristotle, science was *scientia demonstrativa.*

These Latin scholars of the Middle Ages are seen to be commenting on epistemological topics that were of central concern to Aristotle. It is transparent that where he would use *episteme apodeiktike* they used or implied *scientia demonstrativa.* Clearly, therefore, *episteme* can be seen as the Greek equivalent of *scientia.* Feher equates them too, for she contrasts Galileo's astronomy with Aristotle's as mere *opinio* or *doxa* versus *scientia* or *episteme.* Ross's view differs, for he says that "No Greek corresponds to the Latin *scientia.*" Perhaps in a very technical sense this might be true, but insofar as Aristotle was talking about science as demonstrative knowledge, and others were doing the same, *scientia* was equatable with *episteme.*[24] This equivalence together with the quoted excerpts and comments persuade us that there was a continuous tradition from Aristotle to the twelfth century as far as science was both perceived and as it was practiced.

This exploration helps to establish the nature of science as seen by the first great synthesizer; significantly, the linguistic usage by itself gives that nature away. For epistemology has been one of philosophy's major concerns over the centuries and science, which we see was denoted by *episteme,* was inevitably regarded as epistemology and therefore inseparable from philosophy.

So philosophically, because of the demands of syllogistic demonstration, and linguistically, because of its denotation by *epis-*

teme, physics (and science) was clearly nothing more than one preserve of what we would call philosophy.

A major task for this chapter has therefore been completed in that it has been established that science was *scientia.* The next task might be seen as not quite so urgent because we intend to identify when this philosophical nature of science may be perceived to have changed. But this perception of change certainly reinforces the arguments for the original status. It also serves as a bridge to a later discussion about the gradual historical disappearance of this view of science.

Scientia in the Late Middle Ages

Stock, who made a special study of the twelfth century, remarks in the introduction to *Myth and Science in the Twelfth Century* that "Throughout the greater part of the Middle Ages *scientia* referred neither to exact science nor to empirically verifiable fact but to all things knowable. Scientific thought and the language of science were inseparable from mythical modes of explaining how the universe arose and functioned." For mythical modes we may read "philosophies." Indeed Stock records that "the Middle Ages, until the 1140's, had no doubt whatsoever that myths . . . might conceal beneath their surface secret . . . philosophical information."[25] This detailed study of Stock confidently supports the position that up to the middle of the twelfth century science remained *scientia,* except that *scientia* had a little more definition than normally allowed.

Now, from many points of view the twelfth century formed a watershed in the development of science.[26] The recovery and translation of Aristotle's works, the emphasis on the Quadrivium rather than on the Trivium, and the establishment of universities support this judgment. The assimilation of Aristotle's works would have stimulated and broadened an interest in scientific subjects; the revival of the Quadrivium would have reflected this interest; and the universities would have reinforced it. However, Stiefel argues that the recovery of Aristotle's works might have been a hindrance to any questioning that others might have stimulated. But there is evidence of the intellectual stimulation that the recovery of Aristotle's own early works on logic provided for those twelfth-century philosophers who cared, as noted later in this section. There is also evidence that the sciences were being classified in handbooks, encouraged in part by the emphasis that was being placed on the Quadrivium for the study of natural science *qua* science, and not as tools for theology. Of course, the natural homes for the dissemination of such learning were universities, and these as an identifiable institution were just appearing in Europe at the time. Bologna, Paris, Oxford, and Salerno had their origins, although not their establishment, between 1088 and 1173.

Such a background justifies a more detailed consideration of science and *scientia*, especially as science achieved a continuous and accelerating development after this fertile period. Again, Stock, without labeling the twelfth century as spawning a conceptual revolution, discusses the intellectual currents that he identified. One of the currents initiated at that time stemmed from the works of Adelard of Bath, William of Conches, and Thierry of Chartes. This proposition, merely supported by Stock, has actually been propounded and well argued by Stiefel, who argues that these men pursued natural philosophy as a "separate, distinct, and legitimate discipline."[27] She contends that this was a radical change in attitude toward science (or nature) and revealed the first vision of the discipline of science. If this conclusion is accepted, then we may be able to locate the terminus of that first period during which science was *scientia*. We locate, in the middle of the twelfth century, the point at which science gained some individuality—or more precisely, the point at which the possibility of its acquiring individuality, of its being weaned from the apron strings of *scientia*, was first perceived if not yet universally accepted.

Stiefel goes further than Stock by arguing that the new view of science as an independent discipline was a conceptual revolution spawned by those three innovators. However, they were only the major figures among a small elite who had come to realize the wonders of the rational faculty. This group of intellectuals saw themselves as "modern"—a term suggestive of a self-conscious break with the past. Because their concern was with science, as we would call it, this self-consciousness speaks eloquently as well of a new view of science.

However, their revolution did not spread its "infection" very quickly or widely. Stiefel adduces sociological and other factors that prevented this propagation, such as raw prejudice. Thus Adelard despairs of those "who will neither accept this method nor heed my advice—they point their finger at me and impute to me madness."[28] This fate has befallen radical innovators everywhere and always, and what it does for us is to support the contention that a radical change in the view of science was taking place. Despite these obstacles, the works of these authors had a marked effect in the following century; Robert Grosseteste and Roger Bacon, for example, were familiar with them. Of course, by then some universities were well established and Bacon himself testified to the importance of Oxford.

The circumstances of the twelfth-century revolution; the delay in its diffusion; and the fact that Grosseteste and Roger Bacon, in the thirteenth century, are deemed to be the first to take up and operate within this radical tradition have made us link the twelfth and thirteenth centuries as the (diffuse) terminus of the first period during which science was *scientia*. Stiefel's work, as noted, provides the evidence for the terminus, and supports our choice of authors, namely Grosseteste and Bacon, to be saddled with particular significance.

We may be sanguine enough to expect to find philosophical and perhaps linguistic changes in Grosseteste and Bacon. All such changes might be concrete enough for us to perceive, if not to argue for at this time, a transitional period lasting up to the end of the seventeenth century when science did finally cease to be *scientia*.

Grosseteste and Roger Bacon

The philosophers of the period on whom we shall concentrate therefore are Grosseteste and Roger Bacon. Anyone to be considered would, in this era, have picked up the recovered Aristotelian tradition, and Grosseteste was *the* major figure in this regard. Furthermore, he fully articulated an experimental philosophy that, when practiced in later centuries, was to revolutionize science. Bacon, a devotee of Grosseteste, is significant because he was the first to propagandize Grosseteste's method.

It is not surprising that Grosseteste's basic views on science were largely Aristotle's, for Aristotle's massive works were the starting point for twelfth-century philosphers—indeed, the only point of departure. Grosseteste explicitly approved of and used Aristotle's syllogistic method, and accepted the two-way inductive-deductive analysis of scientific knowledge. More to the point, he argued that scientific knowledge of a fact was achieved when we could deduce the fact from prior and better known principles—exactly the same as Aristotle's view. What Groseteste says in Latin is "Demonstrativa scientia est ex veris primus et immediatis prioribus et notioribus." The tool for this purpose was the demonstrative syllogism, for "demonstratio est syllogismus faciens scire,"[29] or it is the demonstrative syllogism that leads to knowledge.

That Grosseteste's view of science is broadly similar to Aristotle's and his use of Latin is similar to Aristotle's use of Greek are clear.[30] There seems to be no separate science, only *scientia demonstrativa,* or *episteme apodeiktike,* as Aristotle had said.

We must note, however, that Grosseteste went beyond his "teacher" Aristotle. Although he agreed with the inductive-deductive method of prosecuting science, he felt that enough had not been done to extract as much as possible from the limited certainty of such mutable material. In the deductive regime of this methodology, particular (novel) deductions could be made from generalized law or principle. Grosseteste was apparently the first to remark on this aspect. More important, he argued that these deductions could be tested by observation and experiment, thereby increasing, with success, the certainty of the general principles. These would be better demonstrated and would therefore, under the Aristotelian criterion, qualify as scientific knowledge. This change of perspective may be interpreted as the effect of the twelfth-century revolution on Grosseteste, already referred to, as well as to the student of genius superseding his master.

In particular, Grosseteste is found referring to "the special demonstrative sciences,"[31] so that whatever the general meaning *scientia* possessed it seems that the demonstrative areas of knowledge were perhaps beginning to be regarded as very special, for the major distinction within the field of knowledge had always been that the best knowledge was demonstrative knowledge. Here we see a further refinement that describes such areas of knowledge as could be submitted to the experimental method as "special demonstrative sciences."

We must also highlight the fact that experimentation is one of the twin pillars of the modern concept of science. Therefore when Grosseteste is found talking about *scientia,* made more "modern" by his own originality, and we discover that he qualifies its long-standing description, we must suspect that even subconsciously some separateness, some individuality of science is being revealed. Whether Grosseteste was a very careful writer or not, we must take this slight hint.

Of course, the case made does not deny that Grosseteste often used *scientia* in the old way: we have argued for this on the basis of one example, but there are several. We normally find such ambivalence in all those who operate at the frontier of ideas, of science, or of art. We shall find it in Roger Bacon as well: there is evidence from Bacon's *Opus Maius* that we can interpret in a positive way. Because Bacon consciously operated and wrote as a rebel, inveighing oftentimes against Aristotle, and against medieval practice; and because Bacon practiced and preached the novel methodology of his master, Grosseteste, he would be expected, from powerful psychological and practical motives, to have a changed view of *scientia.*

In his works Bacon emphasizes the study of foreign languages. In a particular discussion of translations from Latin of what we would now call science, he remarks that it is necessary to know the Science ("debet scire Scientiam") as well as the language. He lamented "Sed solus unus Scivit scientias, ut Lincolniensis episcopus; solus scivit linguas."[32] He is saying that as Boethius was the only one who knew languages, so the Bishop of Lincoln (Grosseteste) was the only one who knew the sciences. Particularity is suggested in the sense that as one might know languages, a particular area of knowledge, one might know *scientias,* another area of knowledge: this juxtaposition of *linguas* and *scientias* is suggestive especially in the context. However, *scientias* can still sustain the old connotations.

Yet, again, the juxtaposition of *scientiam* and *scire,* which are cognates, emphasizes some particularity about *scientiam.* One can know in general, or one might possess general knowledge. On the other hand, one must "know" "something in particular": one would not be expected to "possess knowledge of" or "know" either "knowledges" or "knowledge in general"—hence, some presumed particularity of *scientia.* The same ambiguity, already noted, might be pressed on us but we do not accept it in this instance, especially in the context of Bacon's discussion about what we would call science. This point needs some elaboration.

In the context of his *Opus Maius,* to begin with, Bacon often cites examples of *scientia.* It is impressive but not conclusive that these are predominantly what we would call science: the effect of fires, the freezing water, fracture, magnetism, alchemy, herbal medicine, and the rainbow, to name a few. We must note, however, that this is only one change in the meaning of *scientia* that we are suggesting can be found in Bacon's work: this change relates to the (exclusive) subjects that might, modernly, be call science. Furthermore, we can interpret Bacon to be referring to some organized knowledge relating to what we would call physical science. The other change related to its philosophical status and its methodology are yet to be explored. However, if the already mentioned relationship is accepted, then *scientia* must have begun to mean to Bacon something even more circumscribed than the demonstrative knowledge of Aristotle.

We must bear in mind, in taking this stand, that as far as the activity of science was concerned Bacon did import into it *the* modern elements of mathematization and experimentation. Others had regarded Mathematics as extremely important— Plato, for example. Still others had made a case for experimentation. And Grosseteste himself had given importance to both. However, Bacon began to attack Aristotle's system at its core. One way in which Bacon can be seen to be attacking the Aristotelian hierarchy of values is through his attitude to Logic. He does not attack Logic, the central pillar of Aristotelian science with the same vigor or completeness that Francis Bacon was to display, but he displaced it by Mathematics, which is an accepted pillar of modern science. What Bacon did was to give priority of place to the study of Mathematics, instead of Logic, which was against the current practice of the universities. Furthermore, where even Grosseteste regarded experimentation as a means of achieving demonstration through the syllogism—that is, where he failed to attack the Aristotelian system at its core—Bacon began to do so. For these reasons we do not, finally, regard Grosseteste's work as significant as Bacon's for the case being made. Indeed as Bacon recalls, in relation to experimentation, we must recognize that proof, as reasoning that causes us to know, is to be understood with the proviso that the proof be accompanied by its appropriate experience and is not to be understood by the bare proof. He continues elsewhere, on the same topic, to declare that "This mistress of the speculative sciences is alone able to give us important truths within the confines of the other sciences, which those sciences can learn in no other way. . . . Clear examples . . . can be given; but . . . the man without experience must not seek a reason in order that he might first understand, for he will never have this reason except after experiment. . . . For if a man is without experience that a magnet attracts iron, . . . he will never discover this fact before an experiment."[33]

We can interpret Bacon as making a radical statement about experimentation.[34] Unlike Grosseteste, he was not going to integrate experimentation into the Aristotelian inductive-deductive system to gain a little more certainty. That would have been accepting the system, which he otherwise attacked. It is not more of the same (certainty) that he is aiming at: what his "new science" can provide is something that a philosopher "will never have . . . except after experiment."

Even more cogently, Bacon argues that his experimental science has three prerogatives with respect to the other sciences. First, the other sciences cannot have *particular* or *complete* knowledge without it. Second, it could add *new* knowledge to the sciences, which Aristotle's method could not achieve, even partially: it was certainly not geared to do so. Third, Bacon's method could go even further and *create* new subsciences.[35] We can see from all of this how radically different Bacon's experimental science was from Grosseteste's and certainly from Aristotle's. It embraces a new view of science, and we say this in the full knowledge that it is still different from modern scientific methodology that incorporates experimental science.

Another very important point is that Bacon's Mathematization and Experimentation were to be used in the services of a universal science. In the realization of this aim, he saw Mathematics as the "door and key." As he wrote "There are four great sciences, without which the other sciences cannot be known nor a knowledge of things secured. . . . Of these sciences the gate and key is mathematics."[36]

Furthermore, he amplifies this point in a very definite and significant way by going beyond the mere usefulness of Mathematics and by remarking that "Among all the men of influence in the past, who have flourished under the leadership of Pythagoras with a fine mental grasp, it is an evident fact that no one reaches the summit of perfection in philosophical studies, unless he examines the noble quality of such wisdom with the help of the so-called quadrivium."[37]

We have previously referred to the Trivium and the Quadrivium, which existed from medieval times as the Seven Liberal Arts and which, from our vantage point, clearly formed the basis for Arts and Sciences as they are now known. We remarked on the new emphasis given to the Quadrivium in the twelfth century and the importance that this would have had for a new perception of nature. Indeed, already in the early thirteenth century at the universities, students were being encouraged to go beyond the introductory Trivium and Quadrivium to the more advanced *physica* or natural science.

The Quadrivium, comprising as it did Arithmetic, Music (insofar as it *was* an application of Arithmetic), Geometry, and Astronomy was clearly geared for the modernizing of science, and it is of the utmost significance to find Bacon paying it such attention and compliment at that time. We submit that what he felt that it could do puts a seal, for the first time, on the burgeoning individuality of the sciences. Here was the right man at the right time giving methodological importance to something that had already been institutionalized (in the universities), and almost certainly without its historic significance being recognized; something that was to be regarded, in time, as the *sine qua non* of the sophisticated development of science. When we take these views in conjunction with those on experiment, there does seem to be a case to be made for Roger Bacon and his time as a turning point in the concept of science.

Our case gains independent support from Easton on one point. Thus,[38] he argues for a Joachmite influence on Bacon in which there is seen some concept of the idea of scientific progress. Molland[39] also considers the topic in some detail, the important point of his work for our purposes being his attempt to push the origins of this concept of scientific progress back to the Middle Ages from the accepted time of the sixteenth century as argued by Zilsel and others. He pushes it back to the same twelfth century of William of Conches, whose significance we have already noted.

Thus, although Molland himself regards the evidence in the Quadrivium as fragmentary, there is a *prima facie* case for a twelfth and thirteenth century concept of progress in science that incorporates Conches *and* Roger Bacon. The time and the authors are clearly supportive of our own thesis. There is, moreover, an as yet unstated and important connection with our thesis in that a concept of progress in science presupposes some objective concept of Science—indeed, some concept, however rudimentary, of an independent entity, which is what we have proposed.

We do not deny the general opinion as to the state of knowledge or science in the Middle Ages: no one can deny that reality. As Molland says, "In the Middle Ages the situation was different. *Scientia* was far more definitely a property of individuals, and, if externalised was thought of as contained in a relatively small number of books by established *auctores*.[40]

But it is our contention that for Roger Bacon this was ceasing to apply; indeed, he attacked the constant reference to "established *auctores*." So that Molland's point that the third world of independent scientific knowledge that Popper argues for and that

Molland says did not obtain in the Middle Ages, although true, was ceasing to be true for Roger Bacon.

However, Bacon used *scientia* in the old way, and we have indicated as much. But as a pioneer he could not be expected to be incorrigibly consistent. After all he had been brought up in the oppressive medieval tradition, which, in other areas, he found difficulty in shedding. It was no different in his use of words. It is therefore premature to say that a definite and accepted change in *scientia* occurred with Roger Bacon. Yet there is some evidence that *scientia* for all the reasons we have articulated, began, without consistency, to take on in Roger Bacon some of the modern connotations that gained currency only four centuries later through the work and influence of his namesake Francis Bacon and others.

In any case a trap of which we must beware is the assumption of a smooth change, over the centuries, from the generality of *scientia* to the particularity of science. For we shall find that someone like Ockham who, later in the fourteenth century, made significant contributions to science and philosophy retained a fairly strict Aristotelian classification of sciences, and usage of *scientia,* whereas others held views that were more circumscribed.

"Science" in the Middle Ages

Finally, we take a look at English usage to discover in the translation of *scientia* by science, what connotations were retained; that is whether "science" itself in the late Middle Ages meant knowledge according to Aristotle; or whether it had acquired meanings relative to those specialized areas of knowledge that we have come to associate with the word and that were glimpsed in Roger Bacon, or even whether it simply meant knowledge as is often stated.

For it is the English usage at that time (or French, if we were to consider that language) that will tell us most unambigously what were the current and most basic meaning of the word whose translation became "science." Latin as the language of scholarship would carry both basic and refined meanings, whereas the dependent language would convey only the former meanings.[41]

In Medieval English Chaucer is a good a choice as any. From the *Parlement of Foules* we have him saying

> And out of olde bokes, in good feith,
> Cometh all this newe science that men lere.[42]

"Science" here means no more than "knowledge," which can be learnt about the Africa in which Scipio had sojourned. Its general meaning is even more transparent in the quotation from the *Prioresses Tale,* where "science" takes on the connotation of "learned discourse."

> Lady! thy bountee, thy magnificance,
> Thy vertu, and thy grete humilitee
> Ther may no tonge expresse in no science.[43]

And certainly this point is reinforced in the following excerpt.

> For as that seith Solomon: "who-so that hadde the science to knowe the peynes that been establissed and ordeyned for sinne, he wolde make sorwe". "Thilke science", as seith Seint Augustin, "maketh a man to way-menten in his herte."[44]

Here science could more appropriately be associated with wisdom or knowledge of the wages of sin. Therefore, when finally we come upon "science" in the following quotation, we know despite appearances that it has not our special meaning: "science" here could simply mean areas of knowledge or knowledges.

> Litel Lowis my sone, I have perceived wel by certeyne evidences thyn abilite to lerne sciencez touchings noumbres and proporciouns.[45]

The use of the explanatory gerundive "touchings" supports this last view. Of course, this does not mean that there might not have been examples of usage in which the word meant something rather more circumscribed, and deliberately so: what it does mean is that the *currency* of usage was as yet unmodern.

The range of references among Chaucer's works is wide enough to satisfy us about the consistency of use in this medieval author; but it may be usefully extended by citing other authors and other times. Thus, Sir John Mandeville, writing in the fourteenth century about Cathay, recalls that sitting around the emperor were "many philosophers . . . proved for wise men in many diverse sciences, as of astronomy, nigromancy, geomancy, pyromancy, hydromancy, of augury and of many other sciences," and they had "many . . . manner of instruments after their sciences."[46]

These subjects, apart from astronomy, are all related to magic, to divination by different means, and thus the list attests to the nonmodern usage of the word. Of course, it offends even Aristotle's *scientia* for no syllogistic demonstration is anticipated.

Finally, we find Sir Philip Sidney, in the sixteenth century, defending Poesy by asserting that "Now, therein of all sciences, is our Poet the monarch."[47] This is the clearest example yet of the use of "science" to mean simply (area of) knowledge, for even with Mandeville there is the scent of nature around its use. Thus we see that certainly when the English language acquired its modern identity, "science" was more *scientia* than Science, and even more "knowledge" than *scientia*.

Conclusion

We can conclude therefore that neither in Latin nor in English did *"scientia"* or "science" convey the modern connotations of science; nor was *"scientia"* exclusively used to mean undifferentiated knowledge as some authors have implied. *"Scientia"* had also the connotation of specialized, demonstrative knowledge with which science, as we would call it, was associated; but all that the association achieved was the underlining of the philosophical nature of science. *"Scientia"* implied not just knowledge of things but of their causes as well. That there was no separate word or concept to define science is not surprising for it did not exist as a separate, indentifiable discipline: the linguistic usage merely mirrored the conceptual reality.

Of course, we saw some movement in the thirteenth century when Roger Bacon may be thought to have invested *"scientia"* with new and special significance. But no serious linguistic change was sustained. The failure of a modern view of science to take hold at that time might be facilely explained by the inertia or opposition that new ideas normally meet. Stiefel suggests that Aristotle's works were like a deadweight on the imagination and initiative of the twelfth century; but we have cited evidence of the stimulation that Aristotle's works did have. A more plausible explanation is the lack of an infrastructure to take advantage of any novel and far-reaching insights. Thus, the establishment of universities by the thirteenth century might have been responsible for the perception of change in Roger Bacon; indeed, as we noted before, he made special mention of the significance of

Oxford. However, the change was not sustained. The explanation must be that insofar as science had not yet developed all the apparatus of theory formation and experimentation, the twin pillars of the modern conception of science, and insofar as science, as we know it, did not exist, there was no word for it.

An important point, however, is that in the thirteenth century changes in the connotations of science and *"scientia"* began to be observed, however dimly, and that one can associate this linguistic change with the conceptual revolution initiated in the twelfth century and taken up by Grossesteste and Roger Bacon in the thirteenth. The fact that Bacon himself and his heirs quite often subscribed to the old connotations does not deny the changed perceptions in such a deeply culture-bound concept. In a way one of the best arguments for this case is that in the nineteenth century when everyone would agree that for quite some time science had ceased to be *scientia,* yet even someone like Ruskin could maintain the opposite.

That which we would have called science was still, in the late Middle Ages and up to the Renaissance, an integral part of a philosophical system (Roger Bacon notwithstanding); for science merely helped in the central philosophical concern of epistemology: it was not autonomous. Of course, with this background in view it is perfectly understandable that when, in later centuries, one area of knowledge through its methods showed that it could provide the best approximation to the eternal quest for certain knowledge, that it was the tool for epistemology, then it would appropriate unto itself the very name of Science—*only now with a different, modern meaning. However, it was still performing the same epistemological task.*

The central point being made in this chapter is that, fundamentally, science has a philosophical nature that is often obscured by these conceptual and historical developments. It is our contention that to be aware of the epistemological status of science, both historically and actually, is to have a fuller and more wholesome view of it; and it is only from a knowledge and understanding of the developments we have described that an appreciation of this aspect of science can be achieved.

2

Classical Duality and the Nature of Light

For classical physics has only gradually shaped the notions of particle and wave to logical contraries or opposites: a . . . study of the history of the logical relationship between these two notions, . . . and of their development to the status of fundamentally incompatible and mutually exclusive conceptions still remains a project for future research.

—Max Jammer, *The Conceptual Development of Quantum Mechanics*

Introduction

Whereas it may be argued that the concern of the previous chapter can be described as metascientific, the concern of this chapter is both more patently scientific as well as current. The example chosen to illustrate the philosophical nature of science, the duality of light, has been of active interest among both students and teachers of physics, and many more besides, for centuries; at the same time, it has often stimulated discussions of a philosophical nature. Although the nature of the current scientific debate surrounding the chosen topic might seem to be different from Aristotle's, our historical perspective will reveal that the basic question surrounding the topic has always been philosophical; indeed, it has always been the same philosophical question. What we explore in this chapter therefore is how a particular topic illustrates that its scientific substance is rooted in philosophy—in the same philosophical questions first asked by Aristotle.

In fact, we can trace the concept of duality back to the Pythagoreans among whom it held sway as a pervasive, mystical principle. It has turned up in many different guises since then so that we might think that it has a certain epistemological force.

For example, Descartes in the seventeenth century reduced the world to thought and extension, qualities of the substantial entities of mind and matter; and in modern physics, according to de Broglie, we had the situation in about 1900 that "physics was hence divided into two: physics of matter based on the concept of corpuscles . . . , and secondly radiation physics based on the concept of wave propagation in a hypothetical continuous medium."[1] This clear, ontological distinction did not last for very long. Based on a neo-Pythagoreanism as well as on sound, theoretical foundations, corpuscles and radiation separately assumed dual natures. The duality of radiation, for example, with which Einstein and Compton embarrassed us only provoked the same de Broglie to launch into inspired arguments for the duality of particles as well. Particles were now waves also, and waves particles. The ontology remained, but the distinction did not.

We note, in passing, that this double duality persuades us of some unifying physical entity of which *particle* and *wave* are merely contingent manifestations. This contrasts with the view of classical physics wherein nature was seen as manifesting itself essentially as particle or wave. The important point, however, is that underlying this change of view there has been the acceptance that *wave* and *particle* exhaust the possible manifestations of nature. Increasingly from the seventeenth century but certainly for numerous centuries before, the arguments about the nature of light have usually been couched in terms of this duality. Now, what physical or philosophical principles demand that the two concepts of particle and wave exhaust all physical descriptions of nature? It is the purpose of this chapter to explore the historical (and logical) reasons for the exploitation of this duality in relation to the nature of light in classical physics. The use of the equivalent duality of *matter* and *motion* will be found to be fruitful as it does illuminate the origin of the duality in the medieval preoccupation with *Substantia aut Accidens,* which takes us back to a consideration of the Aristotelian Categories. We are thus thrown back, again, into the philosophical embrace of Aristotle.

Aristotelian Categories

Aristotle enumerated ten necessary and sufficient categories through which men, as perceiving agents, gain cognition of the world. Of these ten categories one was substance and the other nine accidents.[2] They are listed below with accompanying examples:

Category	Example
Substance	Table
Quality	White
Quantity	Two feet long
Relation	Twice
Time	Yesterday
Place	In room
Position	Sitting
State	Armed
Action	To cut
Affection	To be cut

For Aristotle substance was the basis of the world so that if substances did not exist, nothing else would. Accidents existed only insofar as they inhered in substances. Substance, as the ultimate, individual thing, was seen by him to be neither "predicable of a subject nor present in a subject." Thus, to say that "This table is brown" is to exemplify that *this table* is substance: for we can predicate *brown* of *this table* but not *this table* of itself. Furthermore, *brown* can exist without this particular table, whereas this particular table cannot exist without itself. Substance is also distinguished (from accident) in that it has no contrary, and does not admit of variation of degree. The only possible contrary to *this man* is *this nonman,* which does not exist; and one can not be more or less a man: one is either a man or a table. Thus is the concept of substance explicated.[3]

Aristotle went into much more detail in distinguishing accident from substance. The accidental quality, brown, was unlike substance in that it may be present in the body (subject), because color according to Aristotle, required a material basis. Quality also admits of a variation of degree because a table might be more or less brown. Unlike substance, accidents can have contraries, but, more significantly, if one of two contraries is a quality then the other must be a quality also—certainly not a substance, nor any other accident but quality. For the most distinctive mark of a substance was that it retained its integrity while admitting contrary qualities. Substance, among all things, was in this way unique.[4]

There is ground for much discussion and debate here, but we are not really concerned, in this chapter, with justifying but only with employing Aristotle's philosophical framework. The fact is that a clear distinction between substance and accident was assiduously sought for by Aristotle. Whether we agree with his list of categories and his arguments or not, we must both acknowledge his seriousness as well as the fact that he and his heirs for many centuries used the list and the distinctions in the discussion of the nature of light. In crude, lay terms, the distinction resides in the perception that substance is "things" and accidents the properties of things; and we may, therefore, legitimately ask whether light consists of things or of the properties of things.

Light: Substantia aut Accidens?

It has been suggested that Aristotle's murky views on the nature of light incorporated a crude appreciation of its wave aspect. However advanced this may have been, Aristotle's views on light were not the only or the most important ones in ancient times. It is important to examine them nevertheless, to discover whether his arguments were phrased within the context of his own categories, *substantia aut accidens*.

Aristotle believed that light was neither fire nor any kind of body or transformation. He was denying that light was a material substance, and he emphasized that it could not be a body for two bodies could not exist in the same place. Furthermore, because the opposite of light was darkness, light had a contrary; and as Aristotle specifically stated that a mark of substance is that it has no contrary, whereas accidents may have contraries, it is clear that he is searching implicitly for a description of light within his own established framework that distinguishes substance from accident.

This framework became well established within intellectual communities in succeeding generations. Its use flourished in the West even at the time of the greatest inaccessibility of Aristotle's works, and even among anti-Aristotelians. These latter in the early Middle Ages elevated light to the highest rank of (immaterial) reality. Avicenna specifically identified *lux* with God.[5] Space was also elevated by them to this rank so that light and space and God were coterminal. Thus, when Proclus in the sixth century A.D. says that "Space must be either matter or form,"[6] we can translate this into "Light must be either matter or quality" for the Aristotelian tradition had it that quality (an accident) is the actualization of form. Proclus is effectively subscribing to the distinction within *substantia aut accidens.*

There was no uniform view about the nature of light in the Middle Ages. This was hardly unexpected as we moved from the sixth to the fourteenth century; and from Cairo to Oxford. What was, however, common was the obsessive concern, in the discussions about the nature of light, as to whether it was *substantia aut accidens.* In about 1140 Adelard of Bath is found engaged in a dialogue with his nephew as to whether visible breath (or light) is a substance or an accident. His nephew takes the (logical) point of view that "it . . . must necessarily be either substance or accident."[7] Whether it is corporeal substance or accident is not very clear to him. In the first case it is difficult to see how a body could traverse the enormous distance to the stars and back, in an instant. In the second case, accidents are not able to pass to outer things without a subject.

One hundred years later, c. 1240, Bartholomew[8] is letting us into the secret that "Authorities speak diversely on the question whether light is substance or accident." His own view is that light is substance. Because nothing is nobler than *lux* and substance is nobler than accident, then light is substance. We learn from Bartholomew that Damascus, in particular, regarded light as "having no substance of its own," and that Augustine regarded light as being corporeal or substantial. He was clearly well versed in the literature of the subject and himself expended much philosophical energy in discussing the merits of light as *substantia aut accidens.*

When we look at other major figures in optics in the late Middle Ages we find them arguing within the same tradition. Grosseteste, c. 1250, and his pupil Roger Bacon, c. 1270, both regarded light as nonsubstantial: the central preoccupation of these men was found in Arabia also, among Alkindi, Alhazen, Avicenna, and Averroes. Alkindi, for example, asserted that "It is manifest that everything in this world, whether it be substance or accident, produces rays in its own manner."[9] Even in the seventeenth century Grimaldi, the reluctant rediscoverer of the wave nature of light, could not avoid the attraction of this formulation. In his optical treatise *Physicomathesis de Lumine, Coloribus et Iride,* Grimaldi argues in the first section *de luminis substantialitate* and in the second *de luminis accidentali-*

tate. When he gets down to the task of defining the nature of light, he can only ask "Is *lumen* substance or accident?" In his reading of the literature he had found that the widely held opinion favored light as accident. He, himself, took the alternative view of light as substance.[10]

The Aristotelian framework has lasted for almost two thousand years, and not only indirectly but, through the framing of questions about light, directly in terms of *substantia aut accidens.* There was no way, it seems, for us to have escaped the iron necessity of this duality from Aristotle to Grimaldi, and even beyond to Faraday and to Maxwell. With Aristotle the categories were seen to exhaust the descriptions of our means of perception, and division into substance and accident implied also that *tertium non datur.* The scholastics actually said *substantia et accidens dividunt ens:* as used by them each word was the antonym of the other. We can therefore understand the persistence of the question *"Substantia aut accidens?"* because we understand the logic of it also.

The philosophical framework of *substantia aut accidens* provides us with the means of getting at the physical correlatives, matter and motion, in our exploration of the nature of light. This leads us finally to our topical and more particularized duality of particle and wave.

Light: Matter or Motion?

The "translation" from *substantia aut accidens* to matter or motion is what will concern us in this section. This translation is not straightforward. To begin with, medieval orthodoxy, for example, regarded substance to be spiritual as well as corporeal; and criticism has been leveled at the failure to include events under the umbrella of substance: events may indeed have properties. However, as we have remarked before, we are not concerned with such fundamental philosophical issues in this chapter, important as these issues are; we are concerned with the dominant physical view in which substance is seen as things. Furthermore, we are dealing with the physical world so that we should not be concerned with manifestations of the spirit.

But even within the physical world we find the immateriality of substance being postulated. Faraday regarded force to be the sole physical substance: his electromagnetic field consisted of lines of force, and to him these lines were real. This concept, correct or not, was very fruitful in the hands of Maxwell and others. In the present century Einstein and others also have postulated the field as the fundamental physical entity. We shall again briefly mention this aspect of substance; for the present we shall be concerned with material substance.

As for motion we have to contend with the varied views that philosophers held regarding this phenomenon as they struggled over the centuries toward a proper appreciation both of its cause and its status. We need especially to argue for motion as a (compound) Aristotelian, accidental category, so as to complete the translation. We leave this argument, however, for a later section so as not to interrupt the general development of our thesis. We shall therefore, in this section, take on trust that motion is an Aristotelian category.

If we accept the above conditions, then what we need to show is that the arguments about the nature of light had always been conducted within the framework of matter or motion; we need to emphasize also, whenever possible, that this duality was seen as analogous to the more fundamental one of substance or accident.

An introduction to the argument for our thesis in this section is provided by the fact that Democritian natural philosophy was founded on atoms and the void. For these Atomists the void existed so that the atoms could have the possibility of motion. Their physical world was exhaustively described by the void, atoms and motion. As the void was nonbeing what "existed" was (atomic) matter and motion. In this context it would clearly be unavoidable to ask about light, whether it was a particular configuration of matter or some special manifestation of motion. The same question would be raised about electricity and heat by Faraday and Maxwell. Empedocles, the first to delve deeply into the nature of light, regarded it as the emanation, from a source, of "something" that eventually reached the eye. For him light was matter. Now of the probable "structure" of material substance we recognize the continuous and the discrete so that the description of light as a material substance might have to take account of the possible continuity of matter. Aristotle was one who did not subscribe to the view that the basic structure of the world was atomic. It is interesting, therefore, to see how he dealt with the problem of light, in this context.

Aristotle's universe was voidless, a *plenum,* containing bodies that were impenetrable. He rejected the emanation and propagation of material particles as the correct description of light, postulating instead the existence of a potentially transparent medium (like air); and for him light was a particular *state* of the medium, which state the medium acquired all at once. He disagreed with Empedocles that light was matter; he also disagreed that it could be motion because its passage would be observable as it traveled from "the extreme East to the extreme West." As he concluded "the draught upon our powers of belief is too great."[11] For Aristotle light was the accidental state of the continuous medium.[12] However, it was the movement of this medium that caused us to see. Clearly he did not describe light as motion, and yet he could not avoid associating it with motion—that is, the motion or activity of the medium. We sense that the choices were dictated by the particular natural philosophy. For the Atomists the choice of matter or motion, within their view of the world, was an inescapable choice. We can understand also how Aristotle's choice of the state of the medium was consistent with his view of the world as a *plenum*. Even for people holding divergent views as to the fundamental nature of matter, the choice still remained one between (substantial) matter and (accidental) motion or state.

When we look at the work of the medieval Arabs among whom much of seventeenth century discussion about light originated, we find even more concrete examples of substantial light. At the same time that Alkindi was making general statements about substance and accident, as we have seen, he was saying that light rays have the corporeal properties of the bodies they make visible. Alhazen goes further, in analyzing reflection and refraction, by using the analogy of (material) projectiles being reflected from material walls. In relation to the surface of a mirror "if the motion of the arrow be oblique the arrow will be reflected along a line horizontal and of the same obliquity with respect to the mirror and the perpendicular."[13] He goes on to say that "Light has the same nature of being reflected." Alhazen is seen to be choosing a (material) particulate description of light.

We do not have to, indeed we can not, trace all the detailed changes from century to century but only the general tendency within epochs. In this context we can mention briefly the views of Grosseteste and Roger Bacon, two of the leading (English) writers on light of the thirteenth century whose significance has been argued in the previous chapter. Grosseteste regarded light as coterminal with space; for him it was the fundamental corporeal form, continuous throughout the universe and representing the common corporeity of all matter. A reasonable interpretation would have Grosseteste postulating a kind of material ether. On the other hand Roger Bacon argued that light could not be material because it arose from celestial bodies that were incorruptible. Bacon's optical theory described the visibility of objects in terms of the propagation of what he called "species." This was a quality of form and as he said in relation to species "form cannot be separated from the matter in which it is." He then went on to say that "light in the air is not an object, but a species with a weak . . . existence." This emphasizes that for Bacon light was accident and not substance. But what kind of accident? In the same place the assertion is made that species "is not a motion as regards place, but is a propagation."[14] The first thing that came to mind was motion, and it appears to have been an inescapable commitment. When we move into the pivotal seventeenth century, Kepler's views become important as they immediately precede those of the giants—Hooke, Huygens, and Newton. Kepler regarded light as having the property of flowing or of being emitted by its source toward a distant place. Furthermore, he argued that the lines of those emissions were straight and were called rays. It was in removing the possible suspicion that the rays themselves may constitute light that he opposed matter and motion as the possible descriptions of light. "A ray . . . is nothing more than the motion of light. Just as in physical motion the motion is a straight line but the physical entity which moves is a body, so in the case of light the motion is only a line but that which moves is a given surface."[15] The ray is not light as pure motion; the ray is only the direction in which the matter of light (a given surface) does move. Kepler had chosen the material representation of light.

The fundamental nature of the opposition between matter and motion surfaced again in the nineteenth century, after the true nature of light had apparently been discovered, when Faraday and Maxwell contemplated the nature of heat and electricity. Faraday argued that "We could not reason about electricity without thinking of it as a fluid or a vibration or some other existent state or form." There is no doubt here about the matter-motion opposition nor even of the substance-accident duality, for Faraday himself was uncertain "whether heat and electricity are vibrations or substances."[16] Maxwell, later, was not ready to accept that "electricity is either a substance like water or a state of agitation like heat."[17] The central issue of matter or motion remained unresolved.

That these considerations are very much to the point is fully revealed when we realize that Faraday was speaking about imagination in science. In particular he believed that "in the pursuit of physical science, the imagination should be taught to present the subject . . . in all possible and even impossible views."[18] From the first quotation, therefore, we can confidently interpret him as saying that when the imagination is given full rein, it still cannot think of the physical world except in terms of matter or motion.

All the same it has been an assumption throughout that motion is an accidental category. It was not listed as one of Aristotle's ten categories, and we shall therefore have to provide documentation and arguments for its categorization.

Motion as an Aristotelian Category

For about two thousand years up to the seventeenth century there was a major philosophical concern with motion. The associated problem of motion resided in the unresolved dispute as to whether motion was something distinct from that which was moved, a distinguishable form that could impose itself on sensible objects, or whether it was simply a sequential collection of sense impressions of such objects. The long life of the debate attested both to the difficulty of the subject and to its importance also. Indeed it has been said that no physical variable was more central to the preoccupation of the medieval natural philosopher than motion. Almost inevitably the discussion appeared in a vigorous form with Aristotle, and he placed the debate squarely in the context of substance or accident. The medieval opposition between a distinguishable entity and a set of impressions of sensible bodies or as they termed them *fluxus formae* and *forma fluens,* may be seen also as an Aristotelian opposition between substance and accident. For the medievalists, motion was not on the list of the conventional categories but leading philosophers like Ockham and Oresme referred to it as if it were in the category of quality. Ockham is arguing at one point that change in general and motion in particular is not something separate from the known permanent things like substance and place: he says "Every single thing is either substance or quality,"[19] and this must be true of motion also. Ockham went on to argue for motion as quality rather than substance. The nominalism that he revived made him recognize that although motion was named, was a noun, it did not mean that it "existed." Indeed, this philosophical stance led him to attack Aristotle's categories of substance and relation, and thereby make lasting contributions to our understanding of motion. However, these contributions do not concern us here.

What concerns us immediately is that this advance was based upon the Aristotelian legacy, and we shall examine both how Aristotle himself settled the categorization of motion and what justification there was in using motion rather than another (accidental) category to describe light. To Aristotle *change* was the central phenomenon in the study of nature and it embraced decay, growth, and motion; for example; motion was, simply, one kind of change, namely locomotion. What varies from one kind of change to another are the conditions; for motion he perceived these necessarily to be *place, void,* and *time,* and these conditions would seem to locate motion within the Aristotelian categories of time and place. Indeed, Aristotle maintained that "It is always with respect to substance or to quantity or to quality or to place that what changes changes."[20] That is, change takes place with respect to the already delineated categories. As he further asserted "neither will motion . . . have reference to something over and above the things mentioned, for there *is* nothing over and above them."[21] The only conclusion, therefore, was that motion was compounded of more than one category, which its necessary conditions seemed to demand.

In any case this is the legacy that Averroes received when he interpreted Aristotle as saying that "Motion is not found outside the ten categories. . . . Motion exists in the moved thing; but the moved thing exists in more than one category, therefore motion exists in more than one category."[22]

As for the justification of concentrating on motion[23] and not on another category, Aristotle gave the lead in his assertion that "Nature has been defined as a 'principle of motion and change' and it is the subject of our inquiry."[24] A thorough knowledge of nature must therefore incorporate a thorough knowledge of change and motion. This view has been consistently held throughout the ages, certainly up to the nineteenth century. It is true that Cotter,[25] commenting on the tendency for modern scientists to reduce all natural phenomena to matter and motion, argued the medieval case that colors and sound were not mere motions but existed as such. However, Boyle in the seventeenth century reduced the universe to the two principles of matter and motion, both initially created by God. In Descartes, moreover, the only fundamental force or power in the universe was motion. Much later Wundt, in 1866, argued that all the natural causes were causes of motion. As causes are the fundamental metaphysical notion of the universe, so motion's own importance is underlined.

Indeed it was strongly felt in the nineteenth century that no mere chance dictated that mechanics was the basis of every natural phenomenon; this was both necessary and logical. It was recognized that qualitative changes in external objects could be logically described by saying that one object disappeared and another reappeared. However, there was a problem here in that the identity of being and the indestructibility of matter would go overboard—hence, the importance of motion as change. The fundamental role of motion arises from the fact that here "is one single case where an object does change before our eyes and yet still remains the same, and this is the case of motion. . . . We must [therefore] trace every change back to the only conceivable one in which an object remains identical: motion."[26]

If this was generally believed, then one can understand Descartes insisting that light and all the other forces of nature must be a form of movement, and that his first question about light would be "What kind of motion is the action of light and how it is propagated?"[27] An inquiry into the fundamental nature of light demanded a knowledge of its motion, the evident form of change. The association of motion rather than quantity, say, with light was inevitable.

Light: Particle or Wave?

Historically, it can be shown that the particle-wave contention concerning light was consistently used as the particular, physical model of the more general matter-motion duality. This proposition we shall immediately support. We shall then indicate that this was not the only particular duality that was available and finally argue why it was the only one that made physical sense.

Reference has already been made to the analogy between the behavior of light as smooth surfaces, and physical projectiles thrown against a wall. Ptolemy was one of the first to make this comparison which takes light beyond the notion of matter to that of particles. Alhazen was also to adopt a similar model for light. On the other hand, Philoponus held a view of light that was wavelike. This was partially true also of Grosseteste[28] who specifically described the physical mode of the propagation of light by means of a succession of pulses or waves. This model was based on that of the propagation of sound. It was obviously unclear what the nature of light was but the analogies drawn yielded time and again the physical pictures of particles or waves. This wave picture is spelled out clearly by Roger Bacon, a student of Grosseteste.

Bacon developed the concept of species as some kind of spectral image of the objects seen. Their propagation was achieved in such a way that "the species forms a likeness of itself in the second position of the air. Therefore it is not a motion as regards place, but is a propagation multiplied through the different parts of the medium."[29] He saw the species not as undergoing locomotion but as disturbing the medium such that the disturbance was propagated from place to place. The wave model comes readily to mind with this description.

Grimaldi also is of interest to us, for not only did he deny the wave theory when he clearly had the evidence for it but he conceived of light as a "fluid capable of passing . . . also as a wave through a transparent body." He is talking about light as a macroscopic substance moving as a wave. The combination is unusual, and Ronchi[30] regards the theory and its analogy to water waves as out of place. Yet we notice the matter-motion duality as well as the wavelike aspect of the motion. The wave motion that Grimaldi specifically rejected was the propagation of light by secondary spherical actions, thus indicating that the particular kind of wave motion later to be favored by Huygens had been considered before the latter made full use of the concept.

Just before the first full flowering of the wave theory at the hands of Huygens, we had Descartes insisting that light could be nothing but motion. He had access to the ether that he was the first to bring into physics in a practical way, but for him it was the transmission of pressure (disturbance) between the particles of his vortex that was light. Although Descartes's theory could not be set up, symbolically, as a wave theory, it involved the propagation of a disturbance or motion without matter. It had the rudiments of a wave theory.

Even Huygens's theory, of course, was not a proper wave theory: there was no mention of wavelength or frequency. However, the pictorial aspects and its success gained for it the status of the first wave theory that was worked out and not just postulated or described. In his *Treatise on Light,* Huygens remarks that "We have still to consider, in studying the spreading out of these waves, that each particle of the matter in which a wave proceeds not only communicates its motion to the next particle to it."[31] Here we have not only the rudiments, but the actual reference to waves. In the text there are also the pictorial representations. The duality battle was then properly joined in Newton's writings on light when he "suppose[d] light is niether ether, nor its vibrating motion, but something of a different kind [small bodies] propagated from lucid bodies. They . . . suppose . . . an aggregate of various . . . qualities. Others . . . suppose . . . multitudes of unimaginable small and swift corpuscles." Again, for Newton "it can no longer be disputed . . . whether light is a body. For, since colours are the qualities of light, having its rays for their entire and immediate subject, how can we think those rays qualities also unless one quality may be the subject of, and sustain another? which is, in effect, to call it substance."[32] For Newton light was particles in motion, and he arrived at this view through arguments redolent of Aristotle and his medieval commentators.

The two thousand years leading up to Newton labored toward the crystallization and the articulation of the two alternative descriptions of light. The story is well known from here on and so will not be pursued. In any case, it was our intention only to document and explain how there *arose* this particular duality, which gained significance in Newton and elaboration afterward. What we have done so far is to give the details, historically, of the exploitation of the duality, as well as to make a case for, or explain why, the general matter-motion duality was necessary.

The argument has therefore been brought to the crucial point where important discriminations must be made. If light is matter, is it continuous or particulate? If it is motion, is it translational or wavelike? Our remaining task therefore is to argue for the historical choices of particulate over continuous matter, and of wavelike motion over translational. Whether light is, in fact, particulate matter or wavelike motion is not our concern. Now, it is perfectly possible to argue that light is macroscopic matter like the sea, as opposed to particulate matter like pebbles. But there are serious problems with this supposition especially in relation to the all-pervading aspect of the macroscopic matter. Even in Aristotle, among the Stoics, and in Grosseteste where there might be some hint of a continuous material light, we find, in the first case, that it is the *state* of the continuous medium that is described as light; and, in the second case, the Stoic's doctrine of *pneuma* transformed the static continuum of Aristotle into a dynamic one, entailing the wave propagation of light. In the third case, when the ultimate test is made concerning the physical as opposed to the physiological mode of vision, we find that it is effected through a succession of pulses in this universal matter. It seems that if light is material substance, then it must be particulate.

Furthermore it is implicit in the material description of light that the matter is in motion and not static. As far back as the sixth century Philoponus[33] emphasized that light could not possibly be a static phenomenon, and indeed this view is underlined by the Peripatetic doctrine of "potentiality" and "actuality." In Aristotle motion was automatically associated with light, which was regarded as continually making the transition from a state of potentiality to one of actuality. This militates against light as a universal material medium, for the absurdity of its locomotion then embarrasses us. This also explains why such suspicions, as mentioned previously, on closer inspection reveal a disturbance of the medium itself as the representation of light. We must also not lose sight of the fact that even to the particulate description of light motion is conjoined.

Therefore when Newton, for example, says about light that "If it consisted in pression or motion . . . it would bend into the shadow," he is not conjuring up an alternative description of light devoid of motion. When he concluded that "Light was therefore something of a different kind," namely "small bodies emitted everywhere from shining substances,"[34] he is doing two things. He is subscribing to the twofold exhaustion of the description of light, through motion and matter; and he is subscribing to a description of light as moving particles. Pression or motion as the alternative mode, within classical physics, must therefore be motion *without* matter.

Now, if light is motion can this be translational or must it be wavelike motion? The answer must be that it can not be described distinctively as translational motion. It is clear that in the context of classical physics any one region of the material medium of light can be disturbed so that it suffers permanent displacement or locomotion; the neighboring region would in turn be disturbed and suffer locomotion as well; and so on. Although the initial disturbance is seen to undergo, in this way, a simple translational motion throughout the medium, the logical outcome is that in time, the *whole* matter would undergo infinite displacement, so that the medium would no longer be in position to propagate similar disturbances. Furthermore this hypothetical mode would, physically, be hardly any different from that of particles in motion. If we accept that motion (without matter) is a genuine alternative to matter in motion, then only some kind of oscillatory or vibratory motion of the elements of the medium can be conceived, which both allows neighboring regions to convey the disturbance and yet on the average preserves the pervading medium at rest. We must therefore conclude that if light is motion, then it is some kind of wave motion. Note that we are not interested, in this chapter, in whether light is the continuous propagation of longitudinal or transverse vibrations, or whether it is propagated in the more drastic form of pulses. They are philosophically all the same.

We have already seen the logical and the historical reason for the alternative descriptions of light, and we have just gone through the physical justifications for the same. It is only left for us to schematize the logical and historical developments of this "duality" and to make some concluding remarks.

Conclusion

What we can adduce about the unresolved arguments in classical physics that froze the nature of light between particle and wave are that these arguments are fully comprehensible if not inevitable. They are logically, historically, and physically comprehensible: logically because of the twofold substance-accident division existing in the Aristotelian philosophical framework, which stimulated centuries of discussions about light. The actual particle-wave controversy of the seventeenth century is historically accepted as a more particularized version of the matter-motion, substance-accident divisions that informed discussions in earlier times: the seventeenth century arguments had historical precedent. The classical arguments make physical sense as we saw in the previous section where macroscopic matter and pure locomotion were eliminated.

We can schematize both the historical and the broadly philosophical links of our thesis to make it more vivid without making it less true.

The diagram largely speaks for itself. It is operating as an analytical tool upon the most fundamental, intellectual notions of categories, showing how this class gives rise to the two others,

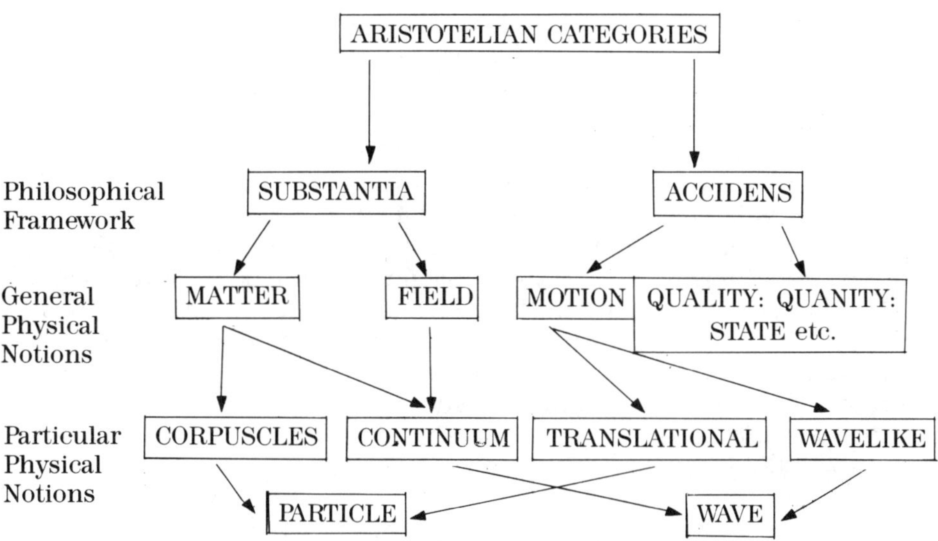

substantia and *accidens. Substantia* is then analyzed for its general physical manifestations, and these are, in turn, analyzed into particular physical components. The same is done for *accidens,* and we arrive at the four possibilities of "corpuscles," "continuum," "translational," and "wavelike." Clearly the first two are mutually exclusive to the description of any particular physical entity; and this is true for the second pair as well. We are therefore left with the four following combinations: corpuscles-translational; continuum-wavelike; corpuscles-wavelike; continuum-translational. The last two are excludable because a corpuscle, although it might oscillate, or undergo wavy locomotion, cannot have a wavelike motion (as opposed to wavy locomotion), which is inapposite. The absurdity of an all-pervading material continuum of light undergoing locomotion has already been pointed out.

We are therefore left, from this analysis, with corpuscular matter in translational motion or wavelike motion in a (macroscopic) medium—that is, with particle or wave.

The schematic arrangement also makes it clear why, although

the exclusive wave-particle dichotomy is comprehensible, it is not inevitable.[35] For it is logically possible from the diagram, as we have seen, to have wavelike motion of particles or translational motion of continuum. The physical arguments eliminate them but pure logic can not. We may therefore say that the wave-particle alternatives are comprehensible although their exclusive role was not inevitable for, logically, there could have been other, additional alternatives.

It is, finally, to be noticed that each alternative, particle or wave, entails *both* matter and motion. For historically light as particles is light as material corpuscles in translational motion; and as wave it is light as the wavelike propagation of a disturbance in a material medium. The difference is that in the first case the emphasis is on matter and in the latter on motion. Perhaps a thought should be given both to the prospect that some of the roots of the modern duality of light can be found here,[36] and to the revelation that a topic of obvious physical interest has roots both in the history of science and in perennial philosophy.

PART II
The Aesthetic Nature of Science

3
Aesthetics in the History of Scientific Theories

Every science begins as philosophy and ends as art.

—Will Durant, *The Story of Philosophy*

Hence it will be useful to develop a field that can fairly be called the aesthetics of science.

—G. Holton, *Thematic Origins of Scientific Thought*

Introduction

Aesthetics or the philosophy of art is an elusive subject, and its association with science even more so; hence, this association has lacked hard-headed formulation. Any putative relationship such as we are proposing is therefore often regarded with scepticism. It is not our intention to feed these reservations, and we achieve our aim by reducing the scope of the association, merely recalling an incontrovertible list of aesthetic elements and discussing to what extent such elements may be discerned, historically, in scientific theories. Aesthetic perception and enjoyment is a matter of taste, and so it is whether the object of our interest is a vase or a scientific theory, as we hope will be revealed in the following pages.

To the extent that the observations above are correct it is illuminating to introduce our subject through three quotations that obliquely exemplify those observations. We do not have any aesthetic *theory* incorporating science.

Science is the response to the demand for information. . . . Art is the response to the demand for entertainment.

—Santayana[1]

63

Every science begins as philosophy and ends as art; it arises in hypothesis and flows into achievement.

—Durant[2]

Science endeavours to proceed from the variety of attributes to the unity of substance.

—Cassirer

Santayana's generalization conjures up the image, first, of a joyless reckoning of nature's score by scientists with short hair and long faces; and second, of a joyful pilgrimage to uncover nature's riches by artists with sanguine expressions and large appetites. This is not a caricature of his assertion but a statement of concurrent twentieth-century views, admittedly largely popular, of the industrial boffin and the bohemian artist, and of their inevitable incompatibility.

It is not too surprising that Santayana took this bleak view of science for he dismissed the utilitarian concept of beauty, believing that "The beautiful does not depend on the useful; it is constituted by the imagination in ignorance and contempt of practical advantage." He furthermore quotes with approval Socrates's famous refutation of the same notion in which the latter argues absurdly that he should be regarded as beautiful because his bulging eyes and large mouth are more efficient for seeing and eating and kissing. It was thus impossible for Santayana to conjoin beauty, the handmaiden of art, and science, which he saw as an entirely practical affair. This view of science is unacceptable for, as we have seen, Aristotelian science had insistent philosophical concerns; and indeed, the pre-Aristotelians' framework for astronomy was largely theological and aesthetic as we shall show. They were passive before the works of nature for they did not believe that they could rifle the notebooks of God.

It was our aim in the previous chapters to show that there is more than a residue of such ancient, "outmoded" forms within our own modern science. We should therefore be expected to adopt the position that Santayana's stark separation of art and science is at the very least historically inaccurate.

Durant's statement is also erroneous, although in a different way, but it is in clear conflict with Santayana's. Although it correctly underscores the dynamic nature of science, it mistakenly implies that science is ever finished. More to the point, unlike Santayana, he clearly associates science with art (and philosophy) and therefore, indirectly, with aesthetics. However, the actual dynamics of scientific change is not as he says for we had, for example, the art of Schrödinger before the philosophy of Bohr. But Durant's errors are either minor or consistent. Thus to the extent that an association exists between art and science and that in art there is a finished product, the painting, the novel or the sonata, whose integrity is often regarded as sacrosanct—to this extent also may Durant have conceived of science as consisting of finished products. This does not mean, however, that we must accept his dross with his gold. As dogma the first two quotations delight us by their brevity and sharpness. As pointed arrows of the wisdom of two philosophers, they penetrate only the suburbs of truth.

Santayana and Durant are neither laymen nor thoroughgoing philosophers of science. Their views, however, are relevant because they are typical, in the deceptiveness of their half-truths, of the whole spectrum from "man-in-the-street" to intellectual. Out of their statements, futhermore, we gain the nugget of the general association of science with art and aesthetics. Cassirer, a professional historian-philosopher, presents us with a more particular and concrete association originating with Leibnitz, a philosopher of science. The distillation of the nature of science contains physical, philosophical as well as aesthetic elements. Thus, substance and attributes are aspects of physical matter as we know it, although the actual names go back to the Aristotelian philosophical discussion of the categories. Furthermore, their particular association is perceived through variety and unity, which we shall posit as two of several aesthetic elements. Leibnitz's position clearly goes beyond Durant's.

We repeat that it is not our purpose to argue for and to establish the logic of the association between science and aesthetics, but only to reveal it, as a fact throughout the historical development of science. Our quotations show a variety of perception and opinion, and they thus exemplify the intractability of the association between aesthetics and science. Our aims are simply to support the implications of those quotations that link aesthetics and science in a general way, and to emphasize some of the aesthetic elements in scientific theories.

Finally, insofar as we have any general aesthetic position, it is not "utilitarian" like Poincaré's. Poincaré maintains that "The useful [mathematical] combinations are precisely the most beautiful."[4] Nor is it like Santayana's position, which asserts that "The beautiful does not depend on the useful; it is constituted by the imagination in ignorance contempt of practical advantage." We combine these two positions, and believe with Schrödinger that the principle of utility can *ultimately* find its own beauty,[5] which is different from arguing that it is usefulness that guarantees beauty. Finally, the positions taken in this and the next chapters are captured by a quotation from Lancelot Whyte that "Aesthetics is no longer an isolated science of beauty; science can no longer neglect aesthetic factors."[6]

Aesthetic Elements

The traditional domain of aesthetics is that of the arts, mainly the visual and aural ones, through which our corresponding senses might be exalted or ravished. But ravishment is properly provoked by the sublime. Even with this qualification, it is difficult to understand how our senses can be ravished by the "agitated" dance of electrons, say, when we can neither see nor hear them. The only concession that many people make to the existence of aesthetics in science is that the one mathematical *solution* is more elegant than another, which often means greater subtlety and concision. Elegance, they might argue is only a sense of style and, anyhow, they might also add, mathematics is not really a science. Therefore, some may doubt whether science can be subjected to aesthetic scrutiny, but this doubt can be allayed by analysis.

Aesthetics, as we have implied, is concerned with the beautiful. An analysis of aesthetic objects like paintings or modern Western music reveals, basically, three components: color (tone), line (rhythm), and form (form). We can make another division into purely formal and purely connotative elments, and it is here that it would seem that *theoretical* science might qualify for (formal) aesthetic consideration. However, with the symbiotic relationship between experiment and theory, it does not make sense to emphasize the theoretical although it will turn out that most of our references are indeed to theoretical science. As it is we shall be discussing, without discrimination, both types of aesthetic elements in the objects of science and shall be making aesthetic judgments on them. We shall be making judgments on their beauty. Now beauty itself is a species of value, and aesthetic judgments are therefore judgments of value. On the contrary, as Santayana argues, intellectual judgments are judgments of fact. The judgment that the quantum mechanics of Schrödinger is more accurate than the classical mechanics of Newton is one of the facts of the case. The judgment that Schrödinger's mechanics is more beautiful than Heisenberg's is one of value. The foregoing sums up more precisely what we are setting out to do, namely the inescapable act of making judgments on the general aesthetic elements in scientific *theories*. We shall incidentally be concerned with elements like unity, variety, symmetry, invariance, form, and simplicity. The more deliberate discussion of formal and connotative elements in *particular* physical theories will be pursued in the next chapter.

In our discussion of "fact" and "value," we might have made a too-sharp distinction between the intellectual and the aesthetic. Thus, some of the aesthetic elements of physical theories, like concision and simplicity, are intellectual qualities also. Such intellectual qualities might be "utilitarian in their origin but aesthetic in their form" and, as Lancelot Whyte further asserts, "The aesthetic sense . . . underlies all human affections, faculties and judgements whether these judgements are seen as aesthetic, intellectual, ethical or practical."[7] Thus, although an element might originate in any category whatever—moral, practical, or intellectual—we can still make aesthetic judgments on it; and we shall have occasion to advert to this opinion in the next chapter also.

Ancient Science

One fears an aesthetic desert in ancient science because it was marginally supported by a mathematics that was rhetorical rather than symbolic. Vieta, for example, invented symbolic algebra in the sixteenth century. This particular fear is supported by Condillac who, in his enthusiastic support for a connection between art and science, argued that both substituted symbols for objects with the difference that scientific symbols are more definite: they seek to achieve perfect, unambiguous expression as well as simplicity, which is an ideal of classical aesthetics. Ancient science, it would seem, lacked the infrastructure for displaying aesthetic elements. But it had geometry of line and figure and, as Crousaz in his *Traité du Beau*[8] argues at length, geometry is the best language for exhibiting the beauty of Nature (Science).

Now the aesthetic of a geometrical formulation may reside in the logical simplicity of the kind of Euclidean system that involves axioms, theorems, and proofs; and this kind of simplicity is not confined to Euclidean geometry only, but embraces even all those non-Euclidean geometries of significance to general relativity. But there is more to it than this: there are the inherently beautiful forms of line and figure—the conic sections. This does not mean that geometrical optics qualifies for our consideration; it does not because the geometry is simply a way of presenting the laws of optics; it is neither an essential, nor a fundamental aspect of the optical phenomena. On the other hand (Cartesian) geometry is an essential part of Newton's gravitational theory whereas geometry is an even more fundamental aspect of Einstein's general relativity. Finally, it is readily appreciated that geometry incorporates the aesthetic elements of symmetry, order, and invariance.

Clearly both Condillac and Crousaz are correct to some extent, but Condillac errs in positing too total an equivalence between art and science. Indeed, even within art, it might be excessive to interpret everything in terms of symbolism. We may take it therefore that aesthetics in ancient science is accessible through geometry. We shall find a path to it also through the symbolism of number.

BIOLOGY

Biological studies, from ancient times up to the scientific revolution of the seventeenth century, were largely classificatory and were of limited use in the understanding of the human body. They were probably no more useful for this purpose than a library's card-indexing system is for understanding education. In the seventeenth century, following Galileo's introduction of quantitative methods in mechanics, quantitative[9] biology and medicine were launched. Even so, because there were no geometrical formulations, anatomy and physiology remain outside our consideration. Old-style natural history would seem to fall into the same category, except that with morphology we are forced to consider elements like symmetry and invariance, and this subject's hold on our attention can indeed be appreciated through the relevant mathematical formalism.

The point is emphasized and clarified by considering beehives, which have obvious aesthetic appeal. We are not concerned, here, with their visible aspect but only with the formal mathematical relations behind these realities. The visible appeal falls within the domain of traditional aesthetics, which is being taken to embrace both art and nature. We are interested in beehives not as art or nature but as science. This distinction is obliquely introduced by von Weizsäcker[10] who vividly recalls his impression of the beauty of the stars, yet questioned whether this "beauty was not somehow in conflict with the fact that those stars are spheres of ionised gas." He decided that the problem could be solved only by seeing "the beauty of the laws governing ionised gases, i.e. by the beauty of the laws of nature." Even so, we shall not pursue the case for Biology here, but will take it up briefly in a later section.

MUSIC

The ancients achieved much in astronomy and music, both of them being elements of the *scientia* of that time.[11] Astronomy and music are of interest furthermore because of the mathematical basis of those achievements, which must be acknowledged whatever their comparative stature as measured against modern science: there is no doubt that the ancients were more ignorant than we. We dismiss the judgment also that religious mysticism impeded rational progress: the Pythagoreans achieved much despite an unmodern blend of faith and reason.

The prime reality to the ancients was their god, a multidimensional one for the pantheistic Ionians and a joyful one for the Pythagoreans. The latter wedded the rationality of their numbers to the Orphic myth, elaborated this compounded religion, and worshiped Orpheus as their god while making advances in *scientia.*

They worshiped number, also, believing number to have some universal reality. Thus Philolaus taught that "All things which can be known have number";[12] and Nicomachus, a neo-Pythagorean taught that "All things that have been arranged by nature according to a workmanlike plan appear as singled out and set in order by Foreknowledge and Reason, which created all according to number."[13] The Pythagoreans thought of number, through aesthetic categories, as the harmonious integration of opposites, that is odd and even. Number, itself an abstraction, nevertheless had for the Pythagoreans a more concrete reality than it does for us. We do no more than hint at this concreteness by noting that with the recording of numbers by dots they equated three and six, for example, with equilateral triangles, and four and nine with squares. Thus the geometry of form and the symbolism of number are combined.[14]

3   4   6   9

It is not surprising therefore that with such a background the Pythagoreans were the ones both to discover the relation between music and number, or the rational basis behind the harmony of strings, and to adopt Orpheus as their god, the same Orpheus who was the god of music. Now they saw their god as a beautiful, harmonious, perfect being. This is true for many who see their god as containing no "opposition between his will and his vision." As Santayana wrote also "It is one of the attributes of God, one of the perfections which we contemplate in our idea of him that there is no duality in him," and, furthermore, "in the contemplation of beauty our faculties of perception have the same perfection as that of god. . . . There is then a real propriety in calling beauty a manifestation of God to the senses."[15]

The Pythagoreans would find acceptable Santayana's view of the similarity of our perceptions of God and of beauty so that God is beauty and beauty God. As Orpheus was a particular form of God, so music was a special manifestation of beauty, and we may, therefore, identify Orpheus with music. Finally, as number was in music, it was in the world because number was in everything: all is number. This particular interrelatedness implies a comprehensively aesthetic view of the world, and we can say of the Pythagoreans that aesthetics, through music, colored their vision of the world. They saw their world through form and number, and it is not too fanciful to see them as early Cartesians. The knowledge at their disposal limited their view of the world almost entirely to the aesthetic. This element would not disappear from later constructions, but certain props and arguments would be discarded. God, for example, would cease to be a necessary hypothesis as the accumulation of knowledge made him into a superfluous and rather inelegant repairman. Thus Napolen's query about the place of God in Laplace's mechanical system received from the latter the reply "I do not need the hypothesis."

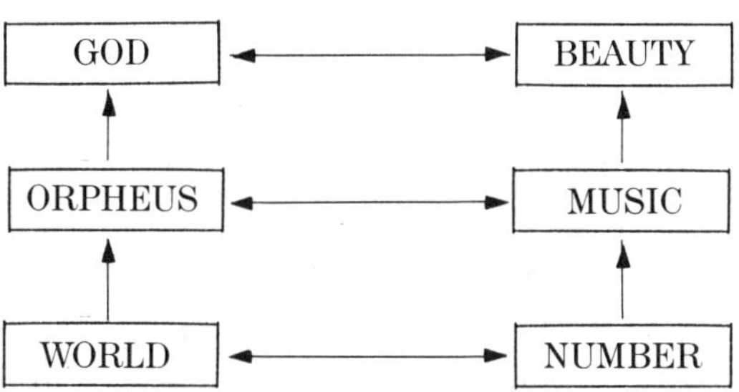

The theological component would disappear altogether but the aesthetic, although acquiring different inspiration, would not.

ASTRONOMY

Plato and Aristotle, whose views on almost everything ruled the intellectual kingdom for centuries, both had a static view of beauty. Beauty, for them, implied repose. The circle had a unique center of repose. The uniqueness for other closed figures disappears—a straight line, so far from possessing a center of repose, by its very definition suggests motion and infinitum. Plato and Aristotle would be among those who, today, might be enraptured by the static symmetry of a vase and yet fail to experience the terrible beauty of a soaring aeroplane. Aristotle was less extreme than Plato, for he did allow reality to change, in the sublunary world of his hierarchic universe, but this was a degraded reality.

For Plato and Aristotle, the problem of motion was an intolerable burden. For Plato, reality was rest and motion was simply a reflection of his Forms in a fitful attempt to reproduce these Forms. For Aristotle, there was an unchanging substrate to which changing or varying qualities could be attached.

This obsession with permanence and stability permeated their whole philosophy, so that Plato's political state is static with no element out of his just place. And insofar as Aristotle was a Parmenidean, he anchored change to an unchanging material substratum. This metaphysical insistence on permanence, on invariance really, has found a larger significance and physical justification in modern theories of relativity and quantum mechanics. Invariance is one of the analyzable elements in aesthetic objects, closely connected with symmetry, as in snow crystals or in Persian carpets. Its role in the aesthetic analysis of modern theories has been no less vital. It is certainly central in Einstein's theories.

When attempts were made by these ancient philosophers to bring the whole universe into their cognitive realm, the aesthetic element, with theological approval, again played a major part. The extension of the Pythagorean philosophy led to the "Harmony of the Spheres."

From their point of view, the earth was at the center of the universe, and the moon, the five then-known planets, and the sun moved in concentric circular orbits around the earth, and below the Empyrean of fixed stars. The orbits were the curved strings of an immaterial, celestial harp, the lengths of these orbits being such that they formed a musical scale. It was believed, for example, that the musical interval between the earth and the moon was one tone, and so on. The Pythagorean scale on this celestial harp was C, D, E♭, G, A, B♭, B, D.[16]

We must notice that this universe had beautiful form—in fact, the perfect form of circles and spheres. Furthermore, this structure was grounded in number and, therefore, in the music of number. Having discovered the surprising relationship between number and music, the Pythagoreans were disinclined to undervalue its significance by limiting its application to mundane matters. We know how precious one's own private discoveries are. In the same way that these discoveries may be unappreciated by others, so the music of the God-created spheres could be heard only by the Master himself.

The circular construction of the Pythagorean universe made appeal to the beauty of the form through its simplicity and perfection. The circle expresses the quintessential harmony of God and of our perception of beauty in that none of its parts are in contest against any others. Santayana's thoughts on the subject probably best sum up the logic of the Pythagorean view, though not necessarily its historical-philosophical basis. He wrote that "This satisfaction of our reason, due to the harmony between our nature and our experience, is partially realized. . . . The sense of beauty is its realization. . . . Beauty therefore seems to be the clearest manifestation of perfection and the best evidence of its possiblity. If perfection is as it should be the ultimate justification of being, we may understand the ground of moral dignity of beauty."[17]

The subsequent shifts in the relative positions of the heavenly bodies did not destroy the aesthetic element. Philolaus made the earth move around some immaterial center; Herakleides made a couple of planets move around the sun, which still moved around a fixed earth. Aristarchus made everything move around the sun with the moon also moving around the earth. But all these motions were circular.

This idea of circularity was first set forth by Plato, but it was Aristotle who made it seem essential; so when Eudoxus and all philosophers through Ptolemy to Copernicus tried to account for observed irregular motions, very elaborate compounded circular motions had to be invented. Epicycles became fashionable, but they still retained the essence of beauty.

Renaissance Science

Meanwhile, with the advent of the Scientific Revolution, ideas about the movement of the planets and their arrangement, motion in general, and space and time had changed from those of the distant, dark night of the ancients. The Aristarchean system of these planets was being established with great hesitations and heart-searching; ellipses were replacing circles as matters of fact; motion, in general, was no longer a property of a body, nor did the particular Aristotelian separation of sublunar from supralunar motion still obtain—the great Newtonian synthesis had done away with that. Motion ceased to be a process; it became a state. The question of the void still raged, with Descartes and Newton being in opposite camps. Where are the aesthetic elements in all these new or modified theories?

First, it must be said that the aesthetic element contained in the concept of circular orbits and a spherical world, or universe, had a tenacious hold upon even the seventeenth-century mind. Copernicus, before, and Galileo, after, the introduction of the Keplerian elliptical orbits, still held fast to the ancient faith. They both accepted a heliocentric universe, but Copernicus forgivably—Galileo unforgivably—refused to be pushed about by the irregular motion that an elliptic orbit would entail. But it was in Kepler's theories, surprisingly, that the aesthetic element was strongest and most persistent. It existed in his view of the structure of the universe, conceived before he discovered the laws of planetary motion. He tried to maintain this view despite the disconcerting light of his almost indisputable laws.

In most ways, Kepler was a new Pythagoras who was enchanted by number and by form. He saw the universe constructed as it was, with only six bodies, out of aesthetic necessity. The six spheres of the planets existed out of the aesthetic necessity to include the (only) five perfect, regular solids: pyramid, cube, octohedron, dodecahedron, and icosahedron. As Koestler wrote, "this idée fixe remained with him all his life and almost all his other work was somehow a means of justifying this concept."[18] This theory fits the terms of reference of this chapter exactly as detailed in the first section, for these regular solids were not regarded by Kepler as substantial but as necessary aesthetic constructs to justify the existence of precisely six planetary spheres. Further clarification of this aesthetic view is achieved by Kepler's explanation that if a cube were made to stand on one of its corners, and an icosahedron on one of its sides, then in both cases the eye would avert its gaze from the ugliness of such a sight. This further aesthetic principle was the criterion used for allotting the solids to their proper places between the spheres. It is interesting that Birkhoff is his *Aesthetic Measure*[19] uses a similar argument to arrive at an aesthetic ranking for various polygons in different orientations.

This was not just a fanciful idea of Kepler's. He included it in his book, *Harmonice Mundi,* one of the most important, systematic expositions of astronomy since Ptolemy. The theory was regarded by a great mind as a serious one, and furthermore, as a neo-Pythagorean, Kepler sought to fit the planetary orbits into a harmonic interval. Whatever we think of the scientific success of the attempt we cannot ignore the aesthetic impulse. It was as a result of this attempt that he first began to consider what relations there might be between planetary period and distance. It led straight to his laws, but that is another matter.

Kepler's theories incorporated the first natural laws, expressing relations between physical quantities. There is no beauty in mere distance or time as physical concepts or measures, although there is some (diluted) kind of beauty in the relation between them. But laws will not qualify as the best that we can do, because they have limited structure. It is to theories that we have to look to find the ultimate aesthetic. We can compare the richer structure of groups with that of sets; and incidentally, it will be found that this is of profound significance in our later discussions. Theories contain within them the oft-stated essence of beauty—namely, unity in diversity: "Quand des Facultez différentes sont toutes également satisfaites, c'est une unité qui charme au milieu de la variété," as Crousaz remarked.[20]

Newton removed the seam between Aristotle's sublunary and

celestial worlds and embraced all motion within the theory of gravitation. All three of Kepler's laws can be deduced from this theory to a high degree of approximation. But we are not seeking for the aesthetic elements in these laws, and it has to be admitted that we cannot fit an eccentric ellipse into the aesthetic framework adopted hitherto. But we can find the elements in the unifying mathematical framework of Newton's theory, where we see the physical phenomena "fitting into place"; for as Agnes Arber, the morphologist, has written, "Rational order (or Uniformity) of Nature is the hypothesis which is tested . . . in all scientific research. . . . This means that the fundamental assumption, the encompassing framework of science, is aesthetic, through the element of form."[21]

Although Newton and most of the revolutionary scientists of his time kept their faith, God was being gradually disposed of, at least as the immediate formative agency. God was for a time the unmoved mover of the celestial system, and its beauty and perfection was simply a reflection of his own: with the complications of Eudoxus and Ptolemy, his cherubims and seraphims had to be delegated some of the manual tasks. However, Newton's theory was now the formative agency, and aesthetic aspects became correspondingly more sophisticated.

So far we have considered some universal phenomena and tried to isolate the aesthetic elements in the formulation of the theories of these phenomena. There is, of course, the still larger framework of spacetime, the ontology and epistemology of which have exercised philosopher-scientists for centuries. Greek philosophers, and certainly Plato and Aristotle, regarded infinity as a measure of imperfection and as a lack of form; their concept of space and time, while subscribing to such an aesthetic, was one of finitude. Aristotle, on this basis, argued the view that there could be no void in space.

Many of the discussions on the nature of space in the sixteenth and seventeenth centuries were what we might call narrowly philosophical. There waas the absurdity reminiscent of the Eleatics, who equated being with identity and were thus led to the impossibility of motion: Descartes equated space with extension and concluded that a void was therefore a "contradictio in adjecto." In any case he felt that "it is repugnant that there should be a vacuum or [a place] where there would be absolutely no thing."[22] In so far as Descartes was aesthetically susceptible, he was, in his theory of the nature of space, not investing it with aesthetic elements, it is true, but at least with aesthetic approval.

It is ironic that it should be Descartes, with his general disregard for aesthetics, who should be the inventor of analytical geometry. For here we have, as in the Pythagorean aesthetic, the conjunction of form and number—only, now, more meaningful and rich, and more comprehensive, too, because of the generalization from arithmetic to algebra. Furthermore, as Cassirer has pointed out, Descartes's achievement epitomized unity in multiplicity for there are for example countless ellipses of various forms that yet are captured in his analytic geometry. Indeed he asserts that the "aesthetic 'unity in multiplicity' of classical theory is modeled after this mathematical unity in multiplicity."[23]

Descartes's theory was one of the achievements of seventeenth-century science in which, as Koyré notes, "the mathematization (geometrization) of nature and therefore the mathematization (geometrization) of science"[24] was predominant. Thus, aesthetic elements, through this formalization, can be found in physics. Form and structure are the essence of geometry and are the essence of beauty, also. But we need, here, to mention counterarguments. As Koyré continues to reflect on the characterization of the revolution of science, he sees that "the *physica coelesti* and *physica terrestris* are identified and unified, in which astronomy and physics become interdependent and united because of their common subjection to geometry. This, in turn, implies the disappearance . . . from scientific thought of all considerations based on value, perfection, harmony . . .: all formal . . . modes of explanation disappear . . . from the new science."[25] Launcelot Whyte makes observations from a similar point of view, remarking that Bacon regarded the form of the thing as its very essence, but that the Cartesian influence had reduced the importance of the formal element, in the sense that any triviality was a mere formality.

At the very moment when nature became more rigorously formalized through Newton's theory of gravitation, and, in fact, when even the Pythagorean mysticism became enshrined in the analytical geometry of Descartes, this very Cartesian influence among others removed (Koyré et al. argue) the necessity of appealing to the formal, aesthetic principle. But this was not so for, although it might no longer have been *formally* necessary to explain these theories by reference to aesthetic principles, the theorists themselves found other reasons, largely *psychological,* for introducing the aesthetic, as we shall argue elsewhere.

The Cartesian influence certainly did not remove the quintessential aesthetic elements from the theories themselves; indeed, these elements were given new life.

Modern Science

PHYSICAL

Descartes's analytical geometry displays the usual aesthetic elements; at the same time it more significantly supports Crousaz and others who have argued that geometry best exhibits the beauty of science, because this "geometry" is so much more deeply incorporated into physical theory than the classical geometry of Euclid. For despite the aesthetic necessity that informed the use of the circle, for example, by the ancients, we are aware, from our position in history, both of the aesthetic aspect of conic sections and of their physical necessity also. All of this has been possible through Descartes's geometry.

When we examine modern physics, we can make an even more substantial case for the aesthetic foundations of science, for this subject can be seen to be based in a fundamental way on geometry or on geometrical formulations. Physics's concern is with the universe, and to the extent that the universe exists it has a large-scale structure and a fine-grained one; and nothing besides of any fundamental nature. Poincaré himself remarked that "It is because simplicity and vastness are both beautiful that we seek . . . simple facts and vast facts; that we take delight, now in following the giant courses of the stars, now in scrutinising with a microscope that prodigious smallness."[26]

We can approach the fundamentals of the universe in the first case through general relativity, and in the second case through elementary particle theory. The substantial case is made therefore through an inspection of the Riemannian geometry, which is the basis of general relativity and through the space-time framework, in which fundamental particle interactions are played out.

However, it is not as if a space-time framework is confined to particle physics for we know that the nature of geometry in physics is of great importance precisely because it leads to an analysis of the space-time system; and it is the space-time structure that is regarded as the basic structure of modern physics. Thus, general relativity is important for our purposes because, at the very least, it puts geometry on "all fours" with the space-time framework, which position is accepted by most philosophers except Anderson, who eccentrically asserts that "Einstein succeeded actually in eliminating geometry from the space-time description of physical systems."[27]

We ignore Anderson's comment and accept that Einstein's aesthetic was a much more comprehensive one than Newton's, with which it would naturally have been compared. However much geometry formed the basis of Newton's analysis, his gravitational theory was not essentially geometrical.

Thus, in the broadest sense, general relativity is aesthetically anchored through its geometrization. We can only add that others support this judgment, although in a general way. Rindler believes that "the most consistent and aesthetic theory of gravitation . . . today is general relativity."[28] This view is shared by Landau and Lifshitz. Others have gone further and suggest that it is only an aesthetic criterion that could decide between Einstein's theory and another.

When we descend to the fine-grained structure of the universe, we come face to face with quantum theory, which is the means by which particle interactions are described. In this regime, it should be recalled both that invariance, symmetry, and the group idea are all liberally used and that geometrization again plays a significant role. We find, too, that the almost mystical Pythagorean union of form and number, which found an operational outlet in the Cartesian method, returns here with all the aesthetic associations. We shall examine these in the next chapter.

Finally, there are the well-known symmetrical and invariant relationships supporting (modern) physics rather than any particular part of it. Thus, the conservation laws of energy and momentum, implying an invariance, are really symmetrical translations in time and space; and the conservation of angular momentum is a reflection of the (symmetrical) isotropy of space. Indeed, for any proposition to be a law it must be invariant with respect to time and space translations as the minimal and necessary conditions.

Thus, aesthetic elements and principles enter into physics from whatever vantage point we wish to view it. From the broadest, most general considerations of structure or of principles to the more sophisticated particularities of interaction and existence even, we seem unable to avoid the aesthetic. We shall make more concrete these general considerations in the next chapter.

BIOLOGICAL

So far, we have considered mainly the physical sciences and, of necessity, only physics. In fact, there are not any large theories in petrology or mineralogy, for example. Most laws that might be worth considering can be subsumed under physics. Even chemical valence, by way of the Kekulé diagram, is seen to be founded on a more primary structure, that of quantum theory. Valence leads us to consider structure as such—from structural inorganic chemistry to morphology in botany and zoology.

Now a pyramid might be beautiful, exhibiting many aesthetic elements, but there is no internal dynamic that impels it to an inevitable manifestation of these elements. On the other hand, there are laws and theories that form the matrix for the beautiful structures of chemistry and biology, and they are therefore of concern to us. We shall briefly consider the use of some of these ideas by D'Arcy Thompson in the field of biology.

D'Arcy Thompson was the first fully to take the obvious step and place biology in a geometrical matrix. In the same way that the relations between states of physical systems can be represented by transformations between, or rotations of, sets of axes, Thompson used coordinate geometry and the operation of transformation from one frame to another to account for the growth and form of different animals, say, within the same genus—hence, the invariance principle of universal applicability. In the same way that motion in the physical Einsteinian matrix takes place along geodesics, growth in the biological matrix takes place along streamlines, or to quote Thompson, "the law of growth which our biological analysis presupposes . . . is one according to which the organism grows or develops along streamlines, which may be defined by a suitable mathematical transformation." He continues by remarking that "The coordinate diagram throws into relief the integral solidarity of the organism, and enables us to see how simple a certain kind of correlation is which had been apt to seem a subtle and a complex thing."[29] He is here taking a holistic or a Gestalt view that has been very important and fertile in psychology and in art.

As a footnote to Thompson's work, we can mention his study on the forms of surfaces and particularly on that of bees' cells, which we mentioned previously. Here the regular beauty arises, as he sees it, from the internal dynamic of economy and simplicity in that this is the only way in which the expenditure of energy in building can be minimized. Minimal energy, as such, takes us outside the realm of aesthetic elements, but it refocuses our attention on the ancients who had their own ideas about the necessary existence of spheres and circles; and in refocusing our attention it emphasizes the continuous and pervasive nature of aesthetic aspects in science.

Conclusion

It is probably wise to deny, explicitly, that this is a prescriptive aesthetic for natural science. The aims of this chapter have been mainly descriptive. Yet it should appear from the presentation that, at the very least, there has been at times a psychological need for such an aesthetic—from the philosophers of Ionia to those of the twentieth century. This was obvious from the general but explicit aesthetic vision of Plato to the more particular one of Kepler. However, the more important fact is that, from the dark and ignorant antechristian days to the present, rhetorical and symbolic theories have both appealed to and displayed many of the formal, aesthetic elements. Psychologically, such an appeal could hardy be stifled even if one could not make a very strong case for it philosophically. If we are to believe Poincaré, then it is not necessary to make any case at all; he believed that the conscious mind becomes aware only of those theories that succeed in passing the scrutiny of what he called the "sensibilité esthétique."[30]

From another point of view, the presence of these aesthetic elements seems to be unavoidable also. In most of the biological and physical theories that are extant, certainly those that we have considered, form enters into them through geometrization and its concomitant aesthetic elements of invariance, symmetry, sequence, order, and structure. Jean Ullmo has considered the philosophical necessity for the presence of these elements. He emphasizes the inevitability of groups, for it is the recognition of them that "enables us to eliminate appearances and to attain the concept of permanent object."[31] The group, in short, provides objectification. At the same time he notices that most mathematical theories arrive at a group as they achieve profundity. So the (mysterious) agreement between mathematical and physical reality is guaranteed by this relationship to groups; and the existence of aesthetic elements in scientific theories is guaranteed by the very nature of groups.

So we see finally that there might be reasons for the use and existence of aesthetic criteria and elements in scientific theories, reasons that can be described as scientific, psychological, and philosophical. Koyré and others have argued that after the seventeenth century there was no longer any scientific need to resort to these elements and criteria. Poincaré's (psychological) position cannot rigorously be argued for: the psychological case may be reduced to a series of plausible assertions to be readily accepted by the converted. Ullmo's philosophical arguments have more substance, but the nature of philosophical argument may enmesh us in endless debate. So no definite case can be made for the need of aesthetic elements in scientific theories.

It is clear nevertheless that these elements have always existed, that scientists and philosophers have always had resort to them, and, most important, that resort to the use of aesthetic elements and criteria have been eminently successful: we shall see more of this in the next chapter. Thus there is a difference between "theory" and fact.

This essay has been an attempt to present the facts of the pervasive influence of aesthetic elements in science without making a case that there might be a theoretical framework within which this influence may be understood. It is worthwhile restating, however, that our philosophical position finds it unproblematic that aesthetics should embrace not only art and nature but science as well.

4
Formal and Connotative Aesthetic Elements in Physical Theories

If one is working from the point of view of getting beauty in
one's equation, one is on a sure line of progress.

—P. A. M. Dirac, "The Evolution of the Physicist's
Picture of Nature"

Introduction

It was our contention in the previous chapter that aesthetics is
no longer an isolated study or criticism of art and nature, for
science can no longer be denied its aesthetic foundations.[1] That
account was mainly a general and narrative one concerned for
the most part with revealing the aesthetic aspects of science,
from the time of the Pythagoreans to the present day. In this
chapter we shall be concerned with describing the nature of the
connotative and formal categories that embrace aesthetic elements, as well as with discussing the role of these elements in
the establishment of some physical theories. The role of such
elements in judgments about theories and in comparative eval-
uations generally will also be examined.

We draw attention to the discussion of aesthetic categories, the
formal and the connotative, for they distinguish between types of
aesthetic elements: where formal elements may be regarded as
some of the available tools in the construction of such theories,
connotative elements may be seen as arising from our interpretation of these theories. Furthermore, although the categories include rather general aesthetic elements like simplicity and elegance, these elements encompass rather more particular ones
like symmetry, which we discussed in the previous chapter. However, whereas our previous concerns were general in that they
were historical and concentrated on Science, our present con-

cerns are particular as they are more analytical and concentrate on three distinct physical theories. Thus, an emphasis on the distinction seems both necessary and natural.

Birkhoff[2] was one of the first to take up the topic of our present concern and treat it in a serious way. In the cited work he explicitly discusses formal and connotative aesthetic elements and suggests, by the way, that in literature the connotative elements might include the whole meaning of a poem. Campbell also relates the meaning of a scientific proposition to an aesthetic element, simplicity, which we shall argue has connotative aspects. He maintains that there are two criteria for the value of scientific propositions, those of truth and meaning, and, furthermore, that a scientific "proposition . . . has meaning in so far as it gives rise to ideas which cause intellectual satisfaction." Campbell than argues that there is a logical dependence of intellectual satisfactoriness on simplicity.[3]

There are two comments that can be made on these positions. Both Birkhoff and Campbell relate the connotative aesthetic to meaning, the one directly and the other indirectly, but Campbell at the same time believes that the "value of a *theory* lies chiefly in its meaning." We shall be concerned with the connotative elements in this chapter where they gain a prominence that they did not have in the previous chapter. Therefore, insofar as we agree with Birkhoff and Campbell concerning the implied distinctions between formal and connotative elements, we have chosen to study these aesthetic elements in *theories,* and more particularly in physical theories.[4] However, we shall first briefly indicate the framework within which all of these elements will be discussed.

Philosophical Framework

The issue surrounding the use of the aesthetic component as a criterion of the successful formulation of scientific theories will be more prominent in this chapter than in the previous one, even though the successes in particle physics were mentioned there. In the previous chapter, we were mostly concerned with the mere existence of aesthetic elements in scientific theories; but the successful use of these elements could not be avoided altogether. In this chapter, we accept the unavoidability and the successes as features to be exploited.

The reference to success might seem to indicate creeping utilitarianism. However, our position, we repeat, is quite different. We earlier referred to Schrödinger's view that the principle of utility can ultimately find its own beauty, and we support and complete this view of the world by adding that conversely the aesthetic can find its own usefulness. Thus there is the aesthetic drive and inspiration of an Einstein whose aesthetic impulses bore fruit not only through the successful formation of his theories of relativity, but also through the successful application of these theories.

There is, however, another important aspect of the aesthetic experience of scientific theories, and that is the response to their aesthetic elements. We have already pointed out that the general nature of such aesthetic response has an intellectual base; however, the specific responses can only be described in terms of normal emotional terminology. Thus, Wittgenstein, in considering aesthetic occasions, suggests that our reactions are always to be taken into account for he believes that perhaps "the most important thing in connection with aesthetics is what may be called aesthetic reactions e.g. discontent, disgust etc."[5] These are rather negative responses. Yet we see that Descartes, in considering what would happen to the walls of a bottle if everything inside the bottle were removed, could only emphasize that "it is manifestly repugnant that they [the walls] be distant, or that there be distance between them, and that, nonetheless, this distance be nothing." For Descartes it is repugnant that there should be a vacuum or a place where there would be absolutely no thing.[6] Even for someone who had no aesthetic theory, we find arguments of aesthetic import at significant points in his science. We find also the same negative referents that we noticed in Wittgenstein, and they lead us to believe that to some extent it is the *absence* of these disturbing elements that induces a satisfying aesthetic response. Of course, someone like Einstein might be expected to provide some support for the philosophical position being articulated, and indeed he was disturbed by asymmetry—the matter-field duality he found "disturbing to every orderly mind."[7]

If we think of the absence of disturbing aesthetic elements as the minimum conditions for an aesthetic response, then the presence of identifiable aesthetic elements might be regarded as sufficient. We shall concentrate mostly on these "sufficient" conditions for the beauty of theories.

Yet, what is the nature of the (basic) intellectual beauty of physical theories? Polanyi has defined the intellectual beauty of a theory as that which causes to be established "a new contact with reality."[8] This definition, agreeably, does not push a utilitarian line but seems to imply in this new contact an element of surprise that is a connotative element, as we shall argue. At the same time, this definition of intellectual beauty is too narrow, but because it contains one of the more important connotative components, it is introduced at this point. Its introduction will allow us to savor the nature of the connotative elements that are to be specifically discussed, in this chapter, in relation to the formal aesthetic elements. We recall the difference between them as residing in the fact that the one is an interpretation of the theory and the other is an entirely objective feature of it.[9]

The distinction that we wish to draw may be exemplified through a comparison of symmetry and grandeur. We are aware of Faraday and Einstein, for example, actively seeking to include the element of symmetry in their theories, and we are aware of the formal yet definite ways in which this element was introduced. Thus, this aesthetic element is an objective feature of those theories; it is there. The surprise of Maxwell's theory or the grandeur of Einstein's refers to our reaction to the results or the implications of these theories and not to their formulation or structure, for there is, indeed, no way that one can legislate for surprise or build it into one's theory. We arrive at the appropriate aesthetic response through our "interpretation" of the theories, and this response might not be universal. Thus Ramsay writes that "where I seem to differ from some of my friends is in attaching little importance to physical size. I don't feel the least humble before the vastness of the heavens."[10] The distinction between formal and connotative can thus be seen to be both real and useful.

Formal and Connotative Aesthetic Categories

There might not be universal agreement about the function of aesthetic elements in scientific theories, but we believe that the following list contains all the major elements whose aesthetic credentials would gain universal assent. They are concision, simplicity, elegance, the element of surprise or wonder, and grandeur.

We shall first look at these elements analytically, the better to define them and to determine which are formal and which are connotative aesthetic elements. We shall then look at them in the context of three specific theories. In everyday discourse, concision is regarded as an admirable virtue both in the one man who expresses his simple idea in a few words, and in the other who might express complicated and profound thoughts with the most ingenious brevity. In the first case we may feel, simply, a sense of rightness. In the second sense, however, we see that there has been accomplished an intellectual feat that provides a certain aesthetic satisfaction. In physics, especially, concision is quite often guaranteed, in both cases, because of symbolic usage. The first case may be labeled as formal whereas the second case often transcends the merely formal.[11]

Simplicity is an element that might easily be confused with concision, and we must be quite clear that we are not reintroducing concision under another name. In fact, the simplicity of a theory might be attainable only through an expansive formulation. A case in point is Einstein's general theory of relativity, which we shall consider in a later section. We shall find simplicity to be a very complex notion, and indeed certain kinds of simplicity will be seen to be partially coincident with other aesthetic elements. It also has both formal and connotative aspects.

This complex element (simplicity) has received a great deal of attention, particularly over the past few decades, but much of this partially successful work has consisted of a detailed analysis of the meaning of the concept and attempts at obtaining measures of it, which emphasis is really different from ours.[12]

We can sharpen our view of the kind of simplicity tht might be relevant by listing the various kinds as analyzed by Bunge. These are syntactical, semantical, epistemological, and pragmatic simplicity; Bunge further subdivides pragmatic simplicity into five types: algorithmic, notational, psychological, experimental, and technical. Even though we use Bunge's terminology, we shall describe some of the aesthetic elements rather differently for we use him mainly as a point of reference.

Syntactical simplicity takes us into the area of grammar rather than prosody—that is, into a consideration of formal rather than connotative elements. Bunge himself calls this the simplicity of forms. We use the analogy of poetry in which "formal" and "connotative" are perspicuously differentiated. There we have the element of meter, which is regarded as formal, whereas the pathos of the poem is connotative. The pathos may originate, partly, in the formal meter but goes beyond it, depending on the individual experience and, therefore, response. It should be clear that one element can be systematized whereas the other cannot, for pathos is just an aspect of the larger meaning of the poem. In science a linear law is syntactically simpler than a nonlinear one.

Semantical simplicity is really the power of a theory, a more general proposition being more powerful than a less general one. Thus, the definition is essentially the same as Bunge's—namely, economy of presuppositions—for a more general proposition naturally economizes, condensing into one statement about phenomena several apparently disparate statements about the same phenomena.[13] The ultimate power, however, resides in the extension of the explanatory and predictive ranges that such a generalization normally entails. Thus, special relativity is semantically simpler than electromagnetic theory. If we remind ourselves that we are talking about the meaning of the theory, then it is clear that we are not dealing with a formal but with a connotative element.

Epistemological simplicity is not always achievable or desirable, for it sometimes leads to shallowness; in fact, physicists often choose terms that are epistemologically complex (and empirically inaccessible), like curved space-time because they lead to semantical simplicity. Epistemological simplicity reduces phenomena to concepts and theories that do not do violence to our common understanding; and, at the same time, it provides us with more certain grounds for our knowledge. In a sense, epistemological simplicity provides us with simpler answers to our questions about the ultimate subject matter of science.[14]

The algorithmic aspect of pragmatic simplicity may be identified with one aspect of the elegance of a theory, for algorithmic simplicity or ease of computation can be seen as a stylistic aspect of the theory. Here we have a formal element, one which does not speak to the (larger) meaning of the theory. The same applies to notational simplicity, which often aids the elegance or the (formal) concision of a theory. It is not an element of fundamental importance.

The psychological aspect of pragmatic simplicity might be regarded as the most intractable because it appears to be the most subjective element. Psychological simplicity or ease of assimilation was already in the eighteenth century being skirted around by Crousaz, who believed that on certain occasions our sentiments are quickened so that we enjoy the elucidation of a difficult idea or proposition. He is arguing within the realm of the connotative aesthetic for he felt that these sentiments could be quickened by grandeur with an element of surprise. There are examples in many areas of physics of this very interaction and response, but we shall not look at any of these. We should notice, however, that Crousaz was discussing the converse of psychological simplicity, namely psychological difficulty whose elucidation induces aesthetic responses nevertheless. We can describe the particular element of psychological simplicity as that element which need not increase our understanding of the world but at least conforms to our intuitive[15] sense of the rightness of a (hypothesis or law or) theory. It includes as well an ability to live in easy harmony with the implications of the theory; and because the subjective figures so prominently in this element, it indeed falls appropriately among connotative elements.

Closely related to the elements concision and simplicity is that of elegance at which we hinted in the previous discussion. Even schoolchildren can appreciate that one mathematical solution is more elegant than another, the former avoiding, among other things, the pedestrian plodding of the other. A theory is a solution to some vexing intellectual problem, and this solution, by the described token, can be more or less elegant than another. Elegance is a sense of style and it involves taste or, according to Rawlins,[16] "the illative sense whose force is intellectual but whose strength is artistic." Elegance is very often, but too narrowly, associated with the pragmatic element of algorithmic simplicity. It almost always implies a subtlety of conception together with a certain grace of exposition and with the soul of wit blown into it. There is no need to emphasize the nature of our aesthetic reactions to such an element as elegance as defined, except to say that it partially coincides with that aspect of concision that transcends the merely formal and is mainly connotative.

The three elements considered so far display some aspect of the formal. The next, the element of surprise or wonder, is entirely connotative and may be found in the reaction experienced in the contemplation of the starry skies, the poetry of T. S. Eliot or the quantum mechanics of Schrödinger. Bronowski[17] explains this reaction by pointing out that "The poem or the discovery exists in two moments of vision: the moment of appreciation as much as that of creation," going on to argue that the great poem and the profound theorem are new to every reader "yet [they] are his own experiences because he himself re-creates them. They are the marks of unity in variety; and in the instant when the mind seizes this for itself, in art or in science, the heart misses a beat." This element should also be emphasized because this essay is addressed not only to an elite band of scientists but to the general educated public also, for whom the element of surprise in the discoveries of science is one of its main extra-scientific elements as well as its most dramatic. Furthermore, the element of surprise is not only experienced once, on first ac-quaintance, but could even increase on greater familiarity with the work when greater appreciation of the antecedents of the theory, of the difficulties of theory making, and of the workings of the theory itself has been achieved. This is not at all surprising, especially when we recall the essentially intellectual aspects of the aesthetics of science.

Grandeur, the last aesthetic element that we shall consider, had already received a passing acknowledgment from Crousaz, and must be posited almost entirely in the external relations of the physical theory. A theory gains its grandeur from its over-powering relation to us, so that it is its awful aspect or its terrible beauty that affects us. We can say that it is the "size" of a theory and the possibility of its being also metaphysically significant to us that are of relevance here. These features, however, reflect its sublimity rather than its beauty, although the sublime is a spe-cies related to that of the beautiful, and a large part of its effect is similar to that of beauty's. What distinguishes it from beauty is greatness. It is the difference between Niagara and a placid stream; it is the difference between a cosmological theory and, say, a theory of superconductivity.

It is useful, at this point, to say a little about the sublime, a part of the aesthetic vocabulary with almost equal significance as the beautiful. However, whereas the beautiful may sometimes be considered to be concerned with order, the sublime may some-times be concerned with disorder; if we associate definite limita-tion with beauty, then the sublime feeds upon enormous propor-tions and infinities. Therefore the sublime is not to be seen as leading to aesthetic appreciation of a different order but a paral-lel appreciation of elements, some of which lie outside the stan-dard vocabulary of the beautiful.[18]

Marjorie Nicholson makes out in her *Science and Imagina-tion*[19] that the concern of the seventeenth-century population began to be centered around the immensity of the universe and that they, like Bruno, "rising on wings sublime," were troubled and enthralled by the prospect. She cites, in particular, George Herbert's verses attesting to the dread of the magnitude of the universe resulting from the new astronomy. Of course, we do not confront the surprise or grandeur of science only through literature but can do so, directly, by reference to our experiences of the theories themselves. There is, however, the drawback that although aesthetic experience is universal, the degree of experience varies, in art as well as in science. All the profounder aesthetic experiences in science are reserved for a minority; and, as Santayana observes, "True test of a work of imagination is the degree and kind of satisfaction it can give to him who appreciates it most." However, the fact that most people, even perhaps most scientists, will fail to gain direct appreciation of the aesthetic aspects of science is no argument against the attempt we are here making. We must not resign ourselves to the attitude that only the finest minds can gain *any* appreciation of the finest things. This chapter attempts to reduce the barriers against such an attitude.

Electromagnetic Theory

Electromagnetic theory was the first major and modern theory in which aesthetic principles were used both in a sophisticated way and creatively. There is that formal aspect of symmetry concerning which arguments were used by Faraday and others to complete the proposition that if certain agents (electric currents) produced certain effects (magnetic fields), then some configuration of these effects should correspondingly produce electric currents. It would not be the last time that the essence of the argument would be fruitfully used. Here we meet the formal element of (physical) symmetry.

The theory is usually presented in the mathematical formulations of Maxwell and Lorentz wherein reside some of the theory's (mathematical) symmetry and concision. We first, and specifically, mention the displacement current, which was original and fruitful with Maxwell and whose origins owed much again to the workings of that formal aesthetic element of symmetry. It is in the mathematical incorporation of this idea that we see the full physical symmetry of the theory embodied in the four Maxwell equations. It was accepted that no (direct) current would "flow" through a dielectric, and yet physical and mathematical symmetry demanded it. This demand proved necessary and fertile.

Philip Morrison also refers to Maxwell's "compact and elegant differential equations,"[20] and only a very slight acquaintance with the theory is necessary to appreciate how much conceptual material is captured by these equations—how concise the theory is. There are two ways in which we can appreciate the element of concision of the theory. We refer, first, to the notations of vector calculus that allow various quantities and their relationship to be represented by four relatively simple equations. We must distinguish this element from that arising from a theory's incorporation into one principle, several principles that had seemed to explain disparate phenomena; this is semantical simplicity. Here, it is the formal linking together, in a concise way, material that is not necessarily disparate, and the element we thus describe is that of (formal) concision.

However, we can also think of Maxwell's four equations as mnemonics, reminding us of many physical laws like Gauss's laws, for example, and others which speak, in detail, to the nature of current, magnetic field, and charge. Here, on the contrary, calculus is used as more than mere symbol of some other physical quantity; so this concision goes beyond the merely formal. It recalls the laws and processes behind the symbols in a way that is alive with meaning, and hence with the connotative aesthetic; and this is achieved through the "divergencies" and the "curls." This formulation is alive with meaning in a way in which, say, Einstein's field equations in the general theory of relativity are not, because the mathematics of the latter is too dense, too opaque. The concision of the latter does not go beyond the merely formal. We suggest furthermore that it is this aspect of concision that goes beyond the merely formal, which Morrison labels as the elegance of the theory; but the theory is elegant only insofar as elegance and concision (partially) overlap, and we would not cite Maxwell's theory as particularly elegant.

Perhaps Maxwell's *forte* was sheer brain power, without felicity, and his often graceless literary exposition excites agreement.[21] In fact, a more direct example of infelicity exists in the inconsistent and impractical mechanical model Maxwell used for his theoretical formulations, a model in which we find "vortices" and other obnoxious physical assumptions—assumptions that were in any case irrelevant to the mathematical formulations and solutions of the theory. There is thus left with us an impression of brute force, a sense of a lack of style.

Electromagnetic theory, however, possesses aesthetic elements of simplicity, but to talk about the simplicity of such a theory might sound like talking to the *cognoscenti*. We need therefore to examine some types of simplicity; we shall not exhaust the list. The epistemological simplicity of Maxwell's theory resides in the emergence of the field. This is a concept and not the whole theory, but it is such an integral part of the theory that we make no distinction between the (scientific) simplicity of theories that we are committed to discuss and that of terms. The field achieves a simpler conceptual framework for electromagnetic phenomena by dispensing with the philosophically obnoxious "action-at-a-distance" on which electrical laws like Coulomb's and Biot and Savart's were founded. Whether Maxwell regarded the field as a separate entity or as the state of a ether is immaterial, for through the postulation of a ether, the dynamics of which explains the phenomena, the grounds of our knowledge of electromagnetic actions are made simpler.[22]

We are talking, here, historically, for it is well known that the theory ran into the epistemological problems concerning the existence of the discrete charge and the continuous field, and concerning the compatibility of the Galilean transformations and electrodynamics. We see from this circumstance that epistemological simplicity, unlike the elegance of a theory, is almost bound to come into conflict with new realities, and reveal how ephemeral it can be.

Another type of simplicity is the pragmatic in which we find five subtypes, one of which, notational, we have already examined; we shall also examine psychological simplicity. It is not immediately clear whether electromagnetic field theory posseses psychological simplicity even relative to the action-at-a-distance version, or absolutely. Because we describe such simplicity as ease of assimilation, the emphasis should be on the psyche rather than on the intellect. It can be argued that even though the theory might be philosophically simpler than the action-at-a-distance description, which it displaced, it could still be difficult psychologically. People have an intuitive understanding of the nature of force and the "need" for contact and will therefore appreciate how the field might push or pull magnets, or how electrified bodies may respond in similar ways. However, the field was more like an occult quality, different from "gravity," which was associated with something concrete and sensible—mass. There were at least psychological blocks against accepting the ether's independent existence throughout space, without either source or support. It is a matter of fact that some people regard God as a suitable epistemological termination or simplification of the problem of infinite regress and yet are unable to treat such a "concept" as psychologically acceptable—that is, as compatible with their own emotions and intuitions. The intellectual "concept" in fact, causes psychological confusion. Electromagnetic field theory brought forth no increase in psychological simplicity, but this was a sacrifice well worth the pain for the power of the theory increased considerably, which is to say that its semantical simplicity increased.

The inherent symmetry of electricity and magnetism that Maxwell's equations embraced reflected the underlying unity and simplicity of apparently disparate phenomena. It is part of the meaning of the theory that there is one electromagnetic field, but it also means that such phenomena could be understood in terms of wave propagation. This connotative aspect of the theory is further enhanced by the subsumption of light under the electromagnetic regime. Of course, when relativity theory increased the semantical simplicity of electromagnetic theory, it did this within a framework of even greater psychological complexities. Not all simplicities are achievable, at least not at the same time.

Finally we examine the purely connotative element of surprise only, as there is no case that can be made in electromagnetic theory for that other element, grandeur. Elements, as we have suggested before, may possess objective existence within a theory while at the same time they elicit a subjective response. The element of surprise must be viewed in this way. Thus, the way in which the velocity of light comes out of electromagnetic theory as a natural but unknown consequence of light being an electromagnetic phenomenon is a prime example of the frisson of surprise that accompanies the unexpected in science. This surprise is not viewed only from a historical point of view for even students learning the theory now in an academic setting experience similar feelings about this quite unexpected power of the theory. The ahistorical aspect of this element speaks eloquently for the intrinsic aesthetic nature of the theory. What we are talking about here is the work done by Hertz not only in producing such waves as were implied by Maxwell's theory but also in causing to be established the connection between light and electromagnetism. A new (and surprising) contact with reality was made. Light had been a mystery for ages, and the nature of its propagation had only recently been satisfactorily determined. Of course, what was being propagated was still a mystery. Now Maxwell's theory, as it was conceived, had nothing to do with light, so that when it was found to provide a respectable theoretical haven for the wave description of light there would have been both satisfaction and delight. When it went even further and, through an interpretation of the solution of Maxwell's equation, provided us with the quiddity that was propagated, there would have been a surprised delight.

It is recognized that Maxwell incorporated into his theory the concept of a unified electromagnetism that Oersted and Faraday had revealed, and this was a major achievement. The existence of waves, although not suspected before Maxwell, was eagerly anticipated after he produced his theory, and the realization of such waves gave sweet satisfaction. However, it was the establishment that light is an electromagnetic phenomenon that induced the feelings of surprise of an intellectual order and of an aesthetic intensity.

Quantum Theory

We examined electromagnetic theory as an example of pre-modern physical theories, and shall examine special and general relativity as examples of modern physics. However, it would be unforgivable to omit mention of that other great modern theory, quantum theory, especially as it is loaded with aesthetic elements. Even so, these elements will get not much more than a brief mention, as we recall the construction of that theory and its influence on particle physics.

First, it is well known how symmetry considerations stimulated deBroglie and Schrödinger in the wave formulation of quantum mechanics; and such considerations were fundamentally physical rather than mathematical. Heisenberg, on the other hand, employed aesthetic elements that were more mathematically formalized. He was inspired by philosophical considerations to use the numbers associated with observable magnitudes, together with formalized algebraic structures, to develop his matrix mechanics. Here physical phenomena emerge from a geometrical framework, the relevant space being Hilbert space in which the vectors are complex, and the space is mapped out by a group of rotations whose aesthetic credentials we have already mentioned.

The role of symmetry is even more striking in particle physics where it is now a methodological tool, pulled out for any attack on a new problem, or any fresh attack on an old one. It plays a central and profound role, certainly in terms of $SU(2)$[23] for example, but also in terms of visible but formal geometrical arrangements. In this context we recall the prediction and subsequent detection of the Ω^- and η° particles largely on the basis of missing elements in such geometrically symmetrical arrangements.[24]

But behind all of these manipulations was the sound PCT theorem, which is a condition of aesthetic proportions on the space-time structure, underlying the many interactions of fundamental particles. We can say, therefore, that fundamentally physics is entirely supported by aesthetic elements. Heisenberg takes this position to its logical conclusion, in arguing against the contemporary search for fundamental particles. He believes that the "fundamental symmetries define the underlying law which determines the spectrum of elementary particles. . . . these symmetries are besides the Lorentz group also SU(2) . . . and the discrete transformations P,C,T."[25] Because of this fact Heisenberg thinks that a decisive change in methodology ought to be effected, and that we should abandon the concept of the fundamental particle and "replace this concept by the concept of a fundamental symmetry." Of course the symmetry of an entity is really its invariance under some transformation.

It is well to recall the P and T invariance originated in geometrical considerations, and although C invariance did not, its origins in Dirac's theory of the electron essentially depended on the symmetrical opposition of matter and antimatter.[26] Yang believes that the "later experimental verification of the existence of the antiparticles constituted . . . one of the most beautiful and forceful demonstrations of the practical consequences of the symmetry principles."[27] We might also compare Dirac's introduction of charge conjugation to the introduction of negative numbers, because of its daring and profundity, and, at the same time, notice the symmetrical opposition of positive and negative, a concept that goes back to Pythagoras.

These thoughts on quantum theory, because of their general nature, could very well have been included in the previous chapter. They are included here because a specific and important comprehensive theory is the object of the discussion. Furthermore, the discussion has revealed in the most concrete way possible how, in relation to the Ω^- particle, the element of symmetry bears fruit.[28]

Relativity Theories

Einstein's theories of relativity provide an excellent display of some of the formal and connotative aesthetic elements that were discussed in previous sections, and they have certainly enjoyed a wealth of reference to their aesthetic aspect and appeal. They command a specially important position in this chapter because Einstein may be said to have had some sort of aesthetic theory; at any rate, he wrote about the role of aesthetic elements in scientific theories. Furthermore, he indicated the ways in which he used aesthetic criteria to establish his theories, or to justify them.[29] In 1901, even before his great papers of 1905, he was writing to a colleague that "It is a magnificent feeling to recognize the unity of a complex of phenomena."[30] When his special theory was first attacked through the experimental work of Kaufmann in 1907, Einstein gave his opinion that he did not believe in Kaufmann's results because they supported theories that were "not explainable in terms of theoretical systems which embrace a greater complex of phenomena."[31] This was an aesthetic criterion, and he was transferring his own feeling about the work he was doing in 1901 into a methodological position in 1907.

The theories that he would support were those in which there was a relation with "the premises of the theory itself, with what may vaguely be characterized as the naturalness or logical simplicity of the premises." Einstein also used the term "inner perfection" to refer to this aspect of theories. This "inner perfection" or beauty captured arch-conservatives like Planck and Wien who suggested that what "speaks for it [the special theory] most of all is the inner consistency . . . one that applies to the totality of physical appearances."[32]

The special theory is dazzling because of the simplicities both informing its construction and resulting from its formulation; and much of these result from formal and connotative aesthetic elements. Thus in the very foundation of the theory we find nothing but two innocuous postulates: no hypothesis about an ether or about mechanics or electromagnetism spoiled the stark simplicity of the core of the theory. There is also the actual simplicity of the individual postulates. Certainly the relativity principle on which Einstein was insisting for all of physics was of such generality as to appear unfruitful, and of such an uncontroversial nature as to appear innocent. The second postulate also shared some of this simplicity in arguing for the (nonunique) independence of the velocity of light with respect to the motion of its source. Thus there is an overriding sense of simplicity in the very foundations of the theory of special relativity.

At the same time, the postulates are loaded with surprise arising from the conjunction of these two scientific propositions that *together* embody a profoundly puzzling contradiction. The surprise is reinforced because it is through this conjunction that Einstein aims to resolve that very contradiction. There is, of course, the liberating surprise when this contradiction is removed, with such ease, by a definition of what it means for two distant events to be simultaneous. The simplicity and the surprise are put in harness,[33] although most of the syntactical simplicity arises from the formal construction of the theory, and most of the surprise from our reactions to the achievements of the theory; it is an aesthetic reaction of surprise that we enjoy also in the "identification of mass with energy."[34]

For many people, of course, the dominant reaction to special relativity is one of great perplexity, which is not conducive to aesthetic or any other appreciation. It has not been easy for many to live with a "time" that is both real and relative and with the mysterious barrier of the velocity of light. Some well-known physicists revealed that they never really understood the theory; others have admitted that they gained insight into it after decades of hard work.[35] Clearly there is no psychological simplicity here, and yet the semantical simplicity of the theory is appreciated even by such people, whose work would have depended on the essential message of the theory as well as on the solutions to problems that only it could achieve. In a sense the most significant of these achievements was the explaining of the Michelson-Morley results, while explaining away the ether. It is an important part of the meaning of the theory, one of its connotations, that the ether is inessential. So this pruning of an awkward and inessential element introduces a semantical simplicity into our apprehension of reality. There is, for example, the semantical simplicity introduced into our understanding of the (relativistic) nature of magnetism and of all those optical phenomena from aberration to the Doppler effect. When we recall the deep intellectual and philosophical perplexity and confusion that these topics induced in the nineteenth century—the kinds of contortions that forced the introduction of Fresnel's dragging coefficient and the Lorentz-Fitzgerald contraction—then the transparent simplicity of Einstein's explanation reflects that intellectual satisfactoriness that we argued was the sign of the aesthetic in scientific theories.

When we move on to Einstein's general theory of relativity we find it again a storehouse of formal and connotative aesthetic elements. Astronomy, space, time, and motion are all included in the formulation of Einstein's theory of General Relativity.

Minkowski first geometrized the special relativistic theory of motion, and this approach motivated Einstein's later generalization. The basic mathematical idea is that a group of continuous rotations map out the whole of the space-time continuum, uniform or accelerated motion being simply reflections of the structure of this continuum according as it is flat, or more or less curved. The equations of motion in general relativity are geometric: they specify geodesics in space-time. Even something as abstract as force becomes, in general relativity, an aspect of the geometry of our now so-called curved space-time, of which matter alone remains independent.

Another important element of Einstein's theory of general relativity is the tensor, a geometric object that entails a fundamental covariance, under group transformations, of the laws that encompass these objects. The tensors themselves stand for a certain invariance that is an aesthetic element of form. Einstein himself wrote about the theory that "scarcely any who fully understands this theory can escape from its magic"[36]—which is strongly suggestive of the "connotative" rather than the "formal aesthetic."

We recall that the inspiration for the theory was aesthetic, in that Einstein wished to generalize within a geometrical framework the Minkowskian formulation of special relativity. The results are the widely inscrutable field equations:

$$(R_{\mu\nu} \equiv) \frac{\partial}{\partial x_\alpha} \Gamma^\alpha_{\mu\nu} + \Gamma^\alpha_{\mu\beta} \Gamma^\beta_{\nu\alpha} = -K(T_{\mu\nu} - \tfrac{1}{2}\, g_{\mu\nu} T)$$ as Einstein produced

them, or

$$G_{\mu\nu} = -8kT_{\mu\nu}$$ as they were subsequently reduced by Foch. T is the energy-momentum tensor; G is the Einstein tensor related to R, the curvature of space-time, and Γ is the metric affinity. Even if these equations are regarded as simple, this is only a formal notational simplicity, much of it invented by Einstein, and on which we are not going to elaborate. In fact the equations conceal mathematics of tedious complexity, much of which is necessary for a full appreciation of the theory. At the same time the theory possesses a well-defined simplicity.

This simplicity resides, not in the mathematical apparatus of non-Euclidean geometry and the accompanying tensor analysis, but in the physical laws, or the physics, resulting from the adoption of this geometry. For, in this choice, we avoid the inordinate complication of having to introduce new laws to describe contracting rods. We avoid, as Carnap puts it, the bizarre consequences of heating a rod—namely, that "all other objects in the cosmos, including the most distant galaxies, immediately contract."[37] The fact is that the complexity of Einstein's theory is mainly algebraic; from a geometric view and in comparison with Newton's theory, Einstein's theory is relatively simple. This syntactical simplicity originating from the epistemologically difficult concept of curved space-time means that we can do without terms like acceleration, and concepts like force and gravity. Force, for example, is now simply the manifestation of the more or less curvedness of space-time. All of this is epistemologically and psychologically very difficult but, as we saw before, not all kinds of simplicity are equally attainable.

We have also, as the centerpiece of this geometric theory, the geodesic hypothesis whose semantical simplicity removes the mystery concerning the equality of inertial and gravitational masses. All bodies, in this theory, follow the geodesic or great "circle," that path of minimum "length," in the non-Euclidean space-time. It applies to a falling stone on earth or to a solar planet. Angel asserts that "few would deny the place of the geodesic hypothesis among the most aesthetically satisfying principles in all of physics."[38] We note that this hypothesis is the Einsteinian equivalent of the "natural" straight line motion of Newton as far as its geometric form goes—which is a formal aesthetic element. It goes further than this, however, in providing us with an understanding of a profound physical fact through a simple hypothesis.

The elegance of the theory is another aspect that has been frequently referred to. Even those who worked closely with Einstein found it difficult to penetrate the method of his thinking. Banesh Hoffman, among others, has reported on this facet of Einstein's life. What we are dealing with, here, is Einstein's style, which, transferred to his theories, we denote as elegance. As Holton remarked, even if we study Einstein's own words as to how he wrestled with theories, we should not deceive ourselves that we "understand" his style, "nor will we forget that other scientists have other styles."[39] What we are considering is something that was individual to Einstein. We argue therefore that the elegance of his theories is individual also.

Elegance, as we have remarked before, is often associated with solutions and might therefore have formal aspects. However, we are not now concerned with these aspects, which relate to the construction of theories, but with the connotative aspects. Dicke asserts that the elegance of the theory resides in the inviolable form that the geometrization gives it; and if we emphasize "inviolable," we are imposing an interpretation on the theory. Dicke is correct both because of the geometrization whose aesthetic force we have often cited, and because of a creative coherence that the geodesic enforces upon a manifold of varying curvature: we accept the theory whole or not at all. However, the elegance of the theory goes beyond this inviolable form. We must bear in mind that Einstein's general theory was not a derivation in the sense that, say, his special theory was,[40] or even Hilbert's own theory of gravitation.[41] Einstein did follow the geometrization introduced by Minkowski, but there was no logical way, however difficult, to the correct field equations; it was neither the case that the field equations were given, and intellectual power had to be expended on solutions and interpretations. Einstein had to pick his way, like an artist, through a mountain of perplexities from the question of appropriate mathematics to the correct choice among many covariant formulations, only some of which could have been discarded on philosophical grounds, and then to the ultimate incorporation of his physical insights into the chosen mathematics. This path was uncharted and impossible to chart so that when Einstein "began to look for a path that restricts the possibilities in a natural manner,"[42] and then took the step so that he arrived at

$$R_{\mu\nu} = -K\,(T_{\mu\nu} - \tfrac{1}{2}\,g_{\mu\nu}T)$$

it could only, ultimately, be *his* style, his own inimitable way of doing things that ensured that "general relativity theory is finally completed as a logical structure."[43]

So we have the elegance of Einstein's general theory for as elegance is often associated with movement, so it is the dynamic aspect of Einstein's mind, reflected in his theory, about which we are talking. We ignore the historical hesitations and look at the actual presented solution where we observe all the intellectual moves as easy, graceful ones, from the *gedanken* experiment in a falling lift, through the natural extension of Minkowski's approach to the insistence on a covariant formulation. This is the fundamental elegance of the theory, and it is emphasized by reference to just one of the examples cited, his *gedanken* experiments that reflect Einstein's penchant for visual thinking, which is an obvious aesthetic mode of the imagination.

Einstein's general theory is being exhibited as the paradigm of beautiful physical theories. Certainly, insofar as the serious study of modern cosmology began with that theory and the element of grandeur, preeminently, resides in cosmological theories, the claim for another aesthetic element in general relativity may be pressed.

The aspect of relativistic cosmology that affects us is that it is concerned not merely with any region whatever in the universe but with the whole universe; in this way it partakes of the grandeur of the universe itself. Mere size may possess no intrinsic merit but such overwhelming size, both physically and psychologically, demands and gets a reaction. Of course, cosmological theory possesses more than a referent to physical size. It can conceivably tell us about our antecedents and our ends. Although these are physical tales they are continuous with the biological ones; and whether we began and shall end with a bang or a whimper is relevant to the grandeur of the theory. Grandeur is exhibited here in the possibility of our history being written on so large a scale.

Poincaré has already been cited as one who reckons, in one of his essays on aesthetics, that "vastness [is so] beautiful that we seek . . . vast facets." Of course, as we have argued, this element partakes more of the sublime, even when we conjoin the physical and psychological grandeur of Einstein's theory to its intellectual grandeur also. As French remarked, "of all Einstein's great scientific achievements, the general theory of relativity is perhaps supreme in its originality and intellectual grandeur.[44] Certainly Einstein had conceived a project of breathtaking ambition and scope, and he reaped the rewards for such daring.[45]

In Chapter 5 we shall have occasion to refer to grandeur again, as an element unlike many others that contributes toward the cultural input of science.

Finally we come to the sense of surprise or of wonder that fills us as we study and begin to understand this general theory. This might read like the unsophisticated response of the ignorant or the uninitiated or, at the very best, of the tyro. Watson accuses popular writers of perpetuating this sense of wonder, and he believes that it is the task of philosophers to ask how much our credulity is imposed upon: 'Credulity is their business.'[46] We would argue, however, that this sense of wonder can properly come only from a sure understanding of the foundations and implications of the theory[47], and is not a function merely of a superficial appreciation of the theory.

Because this element is connotative rather than formal and because it normally has such emotive conntations, we have to be careful not to confuse wonder at Einstein with wonder at his theory. Yet, because the theory is his creation, some blurring of these two categories seems inescapable. Having said that, let us celebrate the abiding sense of awe that we feel that one mind could conceive and carry through what initially seemed such a difficult and unpromising project; that a mind could work at such levels of abstraction and with such faith, mostly alone, to produce such startling results. Perhaps this is one of the most striking aspects of the theory itself. We find also that the sense of wonder of the theory is associated with its grandeur. The theory's grandeur is not only intellectual, but emotional as well, for the realization of our smallness, our insignificance is automatically induced by the scope of the theory, by its seeming cosmological command over our origins and our ends.

Yet there are identifiable elements of the theory that induce such a reaction. Thus, although it was not surprising that a value for the precession of the perihelion of Mercury should have been obtainable once the geodesic hypothesis was accepted, that the value was precisely what observation had shown it to be was in the highest degree astonishing. We remind ourselves, also, that the theory satisfies the criteria for being a good, progressive theory according to the methodological edicts of the leading philosophers of science—in particular, there was not the slightest consideration given, in the formulation of the theory, to the problem of planetary paths or their perturbations. In the language of one particular methodology, this fact was novel. Einstein himself, at the height of his powers and his joy, wrote to his friend Besso after he finally arrived at his field equations "The boldest dreams have been realized. *General* covariance. Perihelion motion of Mercury, *wonderfully exact*" (my italics).[48]

Einstein himself wrote on these matters to his friend Maurice Solovine, expressing surprise that Solovine should

> find it remarkable that the comprehensibility of the world . . . seems to me a wonder or eternal secret. . . . One *should* . . . expect that the world turns out to be lawful only insofar as we intervene to provide order. . . . [However, even] if the axioms of [a] theory are put forward by human agents, the success of such an enterprise does suppose a high degree of order in the objective world, which one has no justification whatever to expect *a priori*. Here lies the sense of "wonder" which increases ever more with the development of our knowledge.[49]

Thus, we have a fitting epitaph for Einstein's work: Einstein's words.

This sense of wonder continues to be experienced not only because new generations of students come fresh to the theory but also because the theory itself continues to grow and to produce astonishing results. Neutron stars and black holes are astonishing in themselves, but they are also solutions of Einstein's field equations. These results were deduced from the theory twenty years after the theory's birth, and the "discovery" of these cosmological objects have reawakened a slumbering interest in the theory while reminding us of its continuing ability to surprise and astonish. That we can understand the universe despite its vast inaccessibility, comprehending its structure and its order through this marvelous theory, will continue to excite us even after Einstein's theory has been superseded.

Conclusion

We have trotted out various opinions on the nature of beauty, for example, that it is related to the useful in one way or another. One such view, an extremely negative one, is espoused by G. H. Hardy, in a discussion of mathematics, the arguments for which being nevertheless relevant to our concerns. Hardy refers to the considerable utility that, say, the differential and integral calculus possess but goes on to state dogmatically that these "parts of mathematics are, on the whole, rather dull; they are just the parts which have least aesthetic value. The 'real' mathematics of . . . Fermat and Gauss and . . . Riemann, is almost wholly 'useless.'"[50] The statement clearly leaves it open to us to infer that it is the useless mathematics (of Riemann, for example) that has the greatest aesthetic value; and we would be the last to underestimate its value. But has it been useless? Hardy's "almost wholly useless" would have left it open to him to avoid criticism by saying that, although Riemann's mathematics may be useful (for Einstein, and physics, and we-know-not-what else), *yet* it is beautiful. It is an exception. Attempting to associate beauty with usefulness in an organic way or within a philosophical system seems an elusive aim, and in any case it is not vital for our interests. Our position, which is essentially one of faith, is that the beautiful will eventually find its own usefulness.

There are, however, other questions that we may ask. Thus, the discovery of elements of beauty in the sample of theories we have examined leads us to ask whether such elements are necessary.

Or are they sufficient. As Dicke says about Einstein's general theory "The intrinsic weakness of a theory based so heavily on . . . such criteria as simplicity and beauty is obvious to all. Yet Einstein was amazingly successful with these techniques and they cannot be ridiculed." Dicke[51] seems to be saying that they are not necessary. Dirac, as we have seen, has actually gone further, saying that "it is more important to have beauty in one's equation than to have them fit experiment. . . . if one is working from the point of view of getting beauty in one's equations, one is on a sure line of progress." Dirac[52] seems to be pointing to their sufficiency. We must, however, balance the picture with Gamow's initial sentiment that if a theory is elegant it must be correct, and his final acknowledgment that, nevertheless, Dirac's elegant theory about the variation of fundamental constants is wrong.[53] So methodological aesthetics in science here seems insufficient, and as always counterinstances are most powerful. We would say that it is not possible to establish a methodology, certainly not of the connotative aesthetic, precisely because we are dealing here with elements that are, by their very nature, not subject to systematization. Thus, even if we ignore, for the time being, Gamow's counterinstance and recall all the favorable cases in which an aesthetic methodology has worked, we still have its intractability to systematization. Perhaps the most that we can hope for is that Poincaré is correct in his judgment that only those theories that pass the scrutiny of what he calls the "aesthetic sensibility" enter the conscious mind. Certainly, the most that can be done is to live always under the shadow of Dirac's ideal and pray constantly for the grace of its blessings.

This might be enough so that Gamow's counterinstance will be of no significance at all. For Poincaré's judgment may simply mean that ultimately Dirac's own theory of the variation of fundamental constants will be found to be correct. The evidence that shows it to be "wrong" might be as inadequate as that which showed to be "wrong" Schrödinger's original and beautiful formulation of a relativistic wave equation, or Weyl's gauge theory of gravitation. Again I believe that the only posture is one of faith. *Faith* alone can be the arbiter in the decision to adopt one theory rather than another because of its aesthetic superiority. There might never be any *reason* in making the choice, unless there was some (nonaesthetic) utilitarian advantage also, like technical or experimental simplicity. Perhaps it is very fortunate that we have had scientists like Einstein and Dirac who did not need *reasons,* who had the faith to believe in the existence of a winnowing aesthetic sensibility.

PART III
The Cultural Nature of Science

5

Science as Method

"[le] trait distinctif de l'époque est . . . l'importance desormais accordée par la publique aux connaissances scientifiques et surtout le generalisation."

—Bernard de Fontenelle, *Textes Choisis*

"Science is not . . . a modification of the black art. . . . It is . . . nothing but trained and organised common sense. . . . The man of Science . . . simply has with scrupulous exactness the methods which we all . . . use carelessly; and the man of business as much avails himself of the scientific method . . . as the veriest bookworm of us all."

—T. Huxley, *Science and Education Essays*

Introduction

The cultural nature of science is supposed to be a topic of enormous range and depth that touches on the historical development of our civilization as well as spans the numerous disciplines of our times. It is a topic that is also concerned with the superficial effects on us of drugs, machines, and electronics, for example; as well as with our (subconscious) metamorphosis in the face of galloping scientific achievements. It is possible, however, to distinguish between science and technology and then to posit that many of the changes are wrought by technology. Such a distinction is valid, and we shall pursue it to the extent of arguing that the most profoundly (but not the only) cultural effect on us all has been methodological; and this may be thought to emanate from science rather than from technology. Our approach, which is largely historical rather than analytical, attempts to take the two-culture controversy set off by Snow and Leavis, more than twenty years ago, further back than others have done before. It is as a by-product of this exercise that the impact of science on our culture is discussed.

In earlier chapters our emphasis on the philosophical nature of science carried the implication that science was not an individual, identifiable entity. For the many centuries when this was so, a confrontation between science and any other part of our culture would have been impossible. By the inverse argument when writers began to bemoan or, at the very least, recognize an incipient conflict between science and letters, say, they must have perceived science as a separate entity. We shall want to argue, later on, that once this identifiable entity arose there was bound to be some jealousy, perhaps even conflict. To the extent that this position is acceptable those writers were correct; they had a clear perception that there were two opposing groups. But whether these groups were cultural entities is another matter. Indeed, we believe that the perception of these writers was shallow.

The historical part of this exercise is necessary because it is not well known that scientists and humanists have been engaged in some kind of territorial dispute whose origin goes back over two hundred years, from the time of Fontenelle to the present day. Most historical accounts take us beyond Snow and Leavis but no further back than to Arnold and Huxley in the nineteenth century when the serious dispute did indeed begin. This territorial dispute, which involved only a few major engagements with decades of attrition and skirmishing, may be seen to have acquired from the original skirmishing a religious flavor also. Preserved Smith, writing about science at the time of the Enlightenment, confirms this assessment metaphorically at least by asserting that "The god of the new religion was Reason; his chosen prophets were a band of writers in France."[1] The great debate between Huxley, Arnold, and others, in the late nineteenth century, came in the wake of that other debate between Huxley and Wilberforce reflecting the religious trauma that Darwin's new theory had induced. In the early 1960s Leavis displayed in his attacks on Snow and others a certain self-righteousness, and assumed in the advocacy of his own case the role of savior of souls. The ascription of religious flavor to these attacks is supported by the fervor of the combatants.

Some of the most futile wars have been religious ones, and most of the verbal fighting between scientists and humanists we believe has been futile, also. Fontenelle felt, to continue the metaphor, that metaphysical and theological arguments were fundamentally fruitless, believing as it was once put that the mind's impatience with endless disquisition is a sure sign of sophistication. We are, therefore, not taking up our cudgels, as many have done before us, in zealous support of science, or literature, or culture. What we hope to do is to study the circumstances of the major confrontations and to make a critical assessment of the views and the motives of the leading protagonists. It will be impossible, from these considerations, not to draw the conclusion that some kind of a "two-culture" concept has always been *perceived* to be "valid" even while each side of the divide fertilized the other. We maintain, nevertheless, that it has been a shallow perception and a false division and that it is so not least of all because it is not possible to make *consistent* claims for science as a separate culture.[2]

The main burdens of this chapter, therefore, are to reveal that this continuing debate is, historically, comprehensible and even inevitable; and to offer the view that science nevertheless, through its methodology, has been an integral part of our culture. Through the arguments that establish this view, we can sharpen our appreciation of the cultural impact of science, even though the arguments and examples are mainly drawn from physics. This approach is largely justified because historically science was mostly what we now call physics.

The Emergence of Scientific Method

The first major and self-conscious attempt to engage in modern science may be attributed to Galileo.[3] It is true that in his analysis of inertia and motion he only adumbrated the true perception of the former, but he introduced his "inclined planes" and a mathematical treatment of the latter. These were the significant changes, taken together, and insofar as we agree with this evaluation it was method that launched science on its modern path. Indeed in Wisan's judgment[4] Galileo was (and still is) the paradigmatic scientist for through the centuries his texts have influenced our views of what science and scientific method should be. However, McMullin demurs.[5]

The first sustained advocacy for science was made by Francis Bacon, especially in his *Novum Organum*. There was no conflict between the humanities and the sciences as we have noted. Bacon took all knowledge for his province, and what he was asking for, at the very beginning of the modern scientific era, was a new method in our intellectual pursuits. Bacon thought that "the entire fabric of human reason which we employ in the inquisition of nature is badly put together." He felt therefore that it was necessary "to commence a total reconstruction . . . of all human knowledge, raised upon proper foundations."[6] Those who had laid down the law of nature before had "done philosophy and the sciences great injury"—for they had "applied no rule, but made everything turn upon hard thinking."[7] It is clear where Bacon was coming from: a regime of no rule—that is, no method—except for the dominance of a purely intellectual hierarchy. It is clear, also, where Bacon was going—straight to a "method, [which] though hard to practice [*sic*], is easy to explain."[8] He was for the most part rejecting mere mental operation and "starting directly from the simple sensuous perception."[9]

Whatever sophistication the nineteenth and twentieth centuries have displayed in distinguishing between scientific and experimental methods, between "experiential" and "experimental," Bacon was writing about physical experimenting in science. In his *Novum Organum* he distinguishes between one method of Cultivating the Sciences and another of Discovering them,[10] and goes on to assert that "The unassisted hand, and the Understanding left to itself, possess but little power."[11] Clearly what Bacon was seeking to establish was a scientific method.

In this light it seems that this method was already permeating the intellectual life of the time. Sypher[12] argues that science and history are culturally linked, and he cites the *Novum Organum* and the *Histoire des histoires* by Bacon and La Popelinière, respectively, to show that, essentially, they conceived of the same scientific method, experiments in the one case and documentation in the other. As far as the early stages of modern science are concerned, we argue that method was the real innovation and that it was not confined to what we would now call natural science. That this was so is further supported by the fact that although Bacon and La Popelinière were both ardent Christians they loosened the bonds between theology on the one hand and science and history on the other, by reducing God as an explanatory mode for the acts of nature; scientific method was taking over.

The Establishment of Science and Its Method

The next stage in the evolution of (physical) science was its establishment consequent upon the successful exploitation of the "novum organum." This stage was set with the publication of Newton's *Principia* in 1687. The succeeding generations of intellectuals, particularly in France but also in England and Germany, perpetuated the method in the movement since labeled "The Enlightenment" so that it affected all human activity. As Preserved Smith writes, "With an intensity of conviction almost religious . . . a chosen band of apostles set out to educate the public in the principles which, they believed, would prove as efficacious in the amelioration of the lot of mankind as they had been effective in explaining the operations of nature."[13] The philosophers intended to employ the scientific method across a wide range of activities. It would become embedded in our culture.

Within the practice of Newtonian science, the method entailed hypotheses, mathematical formulations, and experimentation. Within the wide-ranging activities of the *philosophes,* the essence of the method, as they saw it, was extracted, and a fierce rationalism reigned. It was however mellowed by a constant resort to facts and experience. The first dictionary of the French Academy in the eighteenth century gave among its meanings of *philosophe* that of "a student of the sciences." At the same period, philosopher meant, among other things, one who had "solid principles and above all a good method of understanding facts and drawing legitimate deductions from them."

As one of the leading *philosophes,* Voltaire tried to carry out the program of the Enlightenment. As Cassirer says, "the [arithemtic] ideal of natural science . . . achieves its widest dissemination . . . because [of] Voltaire [who] . . . at once makes this a general question: Newton's method is by no means confined to physics; it holds as well for all human knowledge in general."[14] Voltaire had gone beyond Popelinière in his subscription to the scientific method.

The fact that he intended this wide-ranging application of the method must mean that he did not yet perceive science as challenging the other intellectual disciplines but only as fertilizing them. No antagonisms had yet been conceived or perceived.

Yet science as an autonomous discipline was growing steadily. The crystallization of science as an autonomous discipline occurred in the late eighteenth to the early nineteenth century; but the process had already begun in the early decades of the eighteenth.

Newtonian science caught on, quite naturally, in England. In France it had the Cartesian system to contend with; and it was

Voltaire who was largely responsible for the acceptance of Newton in France. At the same time, Fontenelle was responsible for the spread of the taste for science. Science became fashionable not only within the academic circles of the time but within the drawing rooms also.

The growth of this new empire was seen by some as the first real challenge to the existing literary kingdom. Voltaire, himself, wrote in 1735 that "Verses are not fashionable any longer in Paris. Everyone begins to play the mathematician and the physicist. Everyone wants to reason. Sentiment, imagination and charm are banished. I am not vexed that philosophy is cultivated, but I should not want it to become a tyrant to exclude everything else."[15] This statement is both informative and ironic: it is a fair indication of the rapid inroads that science was making that Voltaire's views should show such a noticeable change over a period of a couple decades. It is ironic in that these changes were largely due to Voltaire.

We need to say three things about this development. First, any person, institution, or discipline that rises into prominence is bound to excite a wariness and perhaps envy and dismay if the entity is seen as a serious challenge. Second, the rise to prominence of science, say, did not automatically spell the impoverishment of its perceived competitors; it did not necessarily imply that two noninteracting subcultures would exist. It could have meant that the existing culture would be enriched. Finally, here with Voltaire we have evidence of the origins of a possible conflict but not the conflict itself. For while he expresses some dismay at the rise to prominence of science, and at the attendant "evils" of this prominence, his wish is not to suppress it or to promote literature but to have both in peaceful and even fruitful coexistence.

Voltaire was one of the leading luminaries of the Enlightenment, and we could take his position on our vexed question as final. However, on this question Fontenelle is at least equally representative of the age. As it was he who greatly popularized science about this time, so it was he who had much to say about culture and the encroachment of science. He was admirably placed to propagate the new science and to comment upon it and its developments, for he lived into his hundredth year, from 1657 to 1757, was thirty at the time of the publication of Newton's *Principia,* and seventy at the time of Newton's death.

He was one of the leading representatives of the new type, *les philosophes,* with the ability to reconcile within himself, as

Maurice Roelens observes, "the demands of the critical spirit and of right thinking, concern about the practical and the useful, and moral and social virtues."[16] Voltaire himself regarded Fontenelle as a man "pour corriger les savants, et pour donner aux ignorants le goût des sciences."[17] Fontenelle had the intellectual and tempermental qualities for the role as spokesman of the age; he had the depth of knowledge and the breadth of sympathy that would seem to qualify him as one of the first great journalist-critics. Cassirer remarks on his abilities in the popularization of science by saying that science "gains acceptance not merely in learned circles but, as a result of Fontenelle's *Conversations on the plurality of the Worlds,* it also becomes a part of the general culture of society." Cassirer adds that "It now goes beyond the sphere of the academics and learned societies. From a mere matter of scholarly interest it now becomes the concern of all civilisation."[18] Fontenelle himself says that the "trait distinctif de l'époque est, sans nul doute, l'importance désormais accordée par le public aux connaissances scientifiques et surtout le generalisation."[19] Thus, Fontenelle goes beyond the recognition of the widespread interest in science to emphasizing that it was generalization, embodied in the scientific activity that was of greatest import.

There was a clear recognition of the existence of two *potentially* hostile groups, of the seeds of conflict. This perception marks the origin of the territorial dispute referred to above. However, neither Voltaire nor Fontenelle saw a stand-off between them; they did not see two cultures standing against each other in mutual distrust and incomprehension. We have already gained a partial view of Voltaire's position. We shall now look more closely at Fontenelle's.

Fontenelle was fully aware of the murky past of the new science that had to assert itself against "the frightening obstacles of authority and power" as it emerged from "the long twilight of barbarism."[20] He was aware, also, that it was the rational faculty and the scientific method that had liberated the mind from its chains of error so that the human condition improved, beyond measure, within a few decades. Fontenelle stressed right thinking, writing that "It is always useful to adopt right thinking, even on useless subjects."[21] At the same time he did stress the practical aspects of science. In his classic plea for science, he discloses the various advantages that have accrued from its study— namely, improved navigation, spectacles, and clocks. He also makes the point that, for historical reasons, could not have been very obvious at the time—that the useful in science is often hidden within the apparently useless. This is just reward for the virtue of right thinking.

Fontenelle seems full of the virtues of science. Did he see any dangers? Now the eighteenth century saw the beginning of modern specialization, in which the sciences assumed individual forms and broke away from the natural philosophy of the previous centuries. This process, together with the great popularity of science, which Fontenelle himself encouraged, made it possible that other intellectual pursuits might suffer at the expense of science. Voltaire, as we saw, noticed that Parisian society preferred to talk science rather than literature. There was no suggestion, however, that scientists and humanists were settling into different camps. If anything it was rather that science was in all the camps, and Fontenelle himself manifested that it was possible to retain all one's human qualities while cultivating the sciences. He believed that the whole person need never be sacrificed to the demands of specialization.[22] For him it was unnecessary for science to dominate the scene—that is, it need not be divorced from the other intellectual pursuits. The method of science was, simply, new flasks for old wine.

It is obvious that it was the methodology of science spanning its rationality and its technology that had entered into the eighteenth-century world. There had been rationaliity before—Descartes's as recently as the seventeenth century—and there had been wheeled vehicles for centuries before Christ. Fontenelle saw things this way also, and characterized the process as the erection of a an edifice by systematic science from the materials brought to it by experimental science.

He gave to method a key role, for it multiplied itself and afforded men the means of appreciating the richer view of things whose richness the method had itself revealed. As he wrote, "Un savant de ce siècle-ci [dix-huitième] contient dix fois un savant du siècle d'Auguste, mais il a eu dix fois plus de commodités pour devenir savant."[23] We can understand from this quotation Fontenelle's appreciation of the cultural importance of the new method.

Fontenelle, in fact, felt that the method of science could be of enormous benefit in areas outside science. He wrote that "The order, the neatness, the precision, the exactness prevailing in good books for some time may well have arisen in that geometrical spirit. . . . A work on ethics politics, or criticism, perhaps even a work of eloquence will be finer, other things being equal, if it is done by the hand of a geometrician."[24] He was not taking a partisan view but showing full appreciation of how science and literature, for example, could complement each other to the undoubted enhancement of culture.

It is now clear that the manifestation of science's triumph took place not only in the superficial context of high-class salons where it could be seen merely as a *succès d'estime;* as a triumph of gossip over truth; of book reviews over books. At the deeper, more fertile levels of exploitation and research, science was entering the culture of the times just as fully, and all the evidence from these two leaders of the period, Voltaire and Fontenelle, suggest that it was scientific method that had triumphed. Indeed D'Alembert, writing after the middle of the eighteenth century, almost a century after Newton's great work, at a time when he would have had a proper perspective on developments, judged that "nearly all . . . fields of knowledge have assumed new forms . . . the discovery and application of a new method of philosophising, the kind of enthusiasm which accompanies discoveries . . . ; all these causes have brought about a lively fermentation of minds."[25]

Prelude to "War"

In the sixteenth and seventeenth centuries the embryonic state of science and the unity of philosophy precluded any confrontation between science and literature, say, or even theology. In the late seventeenth and eighteenth centuries the emergence of science as an identifiable and powerful cultural factor provoked understandable jealousy and resentment. Yet there was no rift. The eighteenth century's concern with science and literature expressed itself in subdued tones. It was a reflection of the nature of the age. Although no separate camps had yet been set up, Fontenelle was clearly very sensitive to developing intellectual currents, as was Voltaire.

Our gradualist perspective suggests that when science fully emerged, when it became established, then would serious conflict arise. We have already mentioned that it was in the nineteenth century that science became fully established; in addition the nineteenth century was a period of great social changes and of cultivated passions. When the above concern arose again it did so in a sharp form and distinct battle formations could be perceived.

It was the Darwinian theory of evolution and natural selection that was the central scientific event of the century, with its incidental iconoclasm and triumphant methodology. Insofar as Newton's *Principia* may be regarded as the major event in the establishment of physical science, Darwin's *Origin of Species* may be seen as leading to the establishment of the biological sciences. Of course, the mere fact of this distinction between physical and biological sciences signifies that there was some larger entity called Science that could be so divided. We shall have more to say about the significance of these divisions in a later chapter.

Many concepts and facts of physical science entered into the consciousness and vocabulary of eighteenth century man; this was true of biological science in the nineteenth century except that the knowledge was more personal and, correspondingly, more traumatic for many. The extent of the trauma resulted not so much from the attempted undermining of deeply held beliefs by some new theory. Humanity had suffered such attacks before and had warded them off. This was an entirely new form of attack in this area, and it seemed that the only alternatives were compromise or isolation; it was the attack of scientific method. There was no failure to recognize this fact, and certainly none concerning its prevalence after the positivism of Comte. As Smith wrote, "In the transvaluation of all values produced by studying . . . nature instead of relying on wisdom of the past and on the common opinion of mankind, revelation faded into mythology and tradition into poetry."[26]

Although 1789 saw the subsidence of the salons, science continued to be popular in France, and the technological benefits made it a tangible reality for many people in Europe. Scientists began to be revered, but science's continued encroachments caused some ambivalence among those who would otherwise support it. Shelley's well-known attachment to science, both at school and afterward, must be placed against his statement in *A Defence of Poetry* that "The cultivation of those sciences which have enlarged the limits of the empire of man over the external world, has, for want of the poetical faculty, . . . circumscribed those of the internal world; and man, having enslaved the elements, remains himself a slave."[27] Again, from the same work, his famous definition emerges that "Poetry is at once the centre and circumference of knowledge; it is that which comprehends all science, and that to which all science must be referred."[28] Science is not dismissed or despised, but it is seen as subservient to poetry. This position goes beyond Voltaire's. Wordsworth's equally famous effusion can be interpreted similarly: "Poetry is the breath and finer spirit of all knowledge; it is the impassioned expression which is the countenance of all Science."[29] This opinion of Wordsworth's was arrived at from the bleak view of science he acquired after attending Davy's lectures at the Royal Institution. Others were even less accommodating. Carlyle, for example, writing in 1831, felt that he could not subscribe to that "progress of Science, which is to destroy Wonder, and in its stead substitute Mensuration."[30] It is clear that there was tension within the intellectual circles of the time. We can see these expressions partly as overreactions to the changes in the material structure of life with stampeding industrialization and the triumph of the mechanistic philosophy. As Arnold wrote in *Culture and Anarchy,* "the whole civilisation is . . . mechanical and external and tends constantly to become more so."[31] Here science was not a separate culture but was destroying culture. He did not like what he saw; besides, it represented a severe challenge to the intellectual status quo: "If friends of physical science were in the morning sunshine of popular favour then, they stand now in its meridian radiance."[32] Here we see clear signs of territorial jealousy where they were only dimly perceived by Voltaire. Furthermore, the latter's conciliatory and unifying approach has been replaced by the aggressive provinciality of Carlyle and Wordsworth. Science had arrived.

Science and Literature: The First "War"

One of the most noticeable features of these skirmishes and battles between science and literature is that it is the literary scholars who are most likely to be antagonistic whereas the scientists are more conciliatory, more aware of the wider possibilities of culture. This fact is consistent with the judgment that the conflicts and battles have arisen largely as a territorial dispute, and out of envy. We have already learned of the almost unanimous antagonism of the men of letters. We shall now examine the views of the scientists and in particular those of Huxley with whom Arnold had the major nineteenth-century battle. Arnold's view will, finally, be put against those of Huxley's.

The extent of the general concern for the problem may be discerned from a perusal of the editorials and articles in *Nature* in 1870, one year after its first issue. As a weekly journal of science it indicated the extent of the practice of science and the general interest in it. But there were anomalies and inadequacies.

Cambridge University had established an examination in natural science in 1852, but in 1870 there was no such faculty at University College, London. It was only then recognizing the special needs of science and in its prospectus for that year it said, among other things, that "The Faculty of Arts in University College, instituted to give a general training in literature and science . . . not only has . . . contained chairs in which scientific instruction has been developed far beyond the needs of arts students, but . . . has come to include others bearing no relation to an arts curriculum." A separate faculty was seen to be an urgent requirement. The editorial in *Nature* commented on the prospectus and indicated that the political motive behind the change was to assure the "independence and dignity associated with the academic title of Faculty." The recognition of separate needs led the university to express the sanguine principles that "science should be cultivated for its own sake," and also that "the pursuit of science should not be divorced from literary culture."[33] *Nature* brought up the issue that Huxley was to take up also, and in the same sympathetic way. We note that the two "principles" locate the central concern of Snow in 1959 and Fontenelle in 1702.

Huxley was one of the leading biologists of his time and like Fontenelle before him he was an accomplished author. As Fontenelle made Newtonianism popular, so did Huxley break down the barriers to Darwinism and made it acceptable. He was a man of broad interests and sympathies and, in matters of faith, a skeptic. He coined the word "agnostic" to describe an old and continuing condition. Huxley was aware of the historical preeminence of the classics and of literature generally. Access to major thought was initially achieved through the classics, which in his century had been challenged by modern literature. He saw science as a new element, and observed that from 1850 the battle between ancient and modern literature had been joined by the physical sciences. He saw scientists as a "guerrilla force . . . of irregulars," and himself as a "full private . . . not devoid of interest." So Huxley saw the battle joined even before Darwin's thesis was published; its publication only intensified the conflict. For Huxley, science was simply *"trained and organised common sense,"* which training could inculcate the most admirable precision. Furthermore, Huxley took the view that "for the purpose of attaining real culture, an exclusively scientific education is at least as effectual as an exclusively literary education," denying that the classicists were "in charge of the ark of culture." For him, "perfect culture should supply a complete theory of life, based upon a clear knowledge alike of its possibilities and its limitations."[34]

Huxley maintained, however, that "there really is no question as to the necessity of purely scientific discipline"; and that there could be no doubt "as to the desirableness of a wider culture than that yielded by science alone."[35] Here we have a balanced attitude based on the perception that method was the crucial inheritance of science, which was poised to enrich our culture. Arnold was rather different.

Arnold was probably the chief conscience of his age, and as far as "culture" is concerned we might say that he had both practical experience and theoretical expertise. He was for thirty-five years an indefatigable inspector of schools in England. One of his editors, Dover Wilson, says of him that he possessed "an unusual combination of qualities and interests. . . . As a poet and a critic he was the most considerable literary figure of the mid-Victorian period . . . as an educationalist . . . he was far ahead of his contemporaries."[36] Arnold achieved the power and the position to which the intellectuals of the ages aspire.

His experience was almost exclusively literary, and he defended literature as the chief element in an academic education and in the perpetuation of culture. In fact, his attitude toward culture was rather bookish. Science insofar as it was literature was included in the cultural "text"; but he denied in "Literature and Science" that it was necessary for a proper pursuit of culture to engage in all those *processes* by which scientific results could be achieved.

This attitude is not as unsympathetic as it appears, and Huxley realized this. Although Arnold excludes experimentation in science from the concept of culture, he does not exclude *knowledge* of science; nor, more important, does he exclude scientific *method.* He denies that by literature he means *belles lettres;* he means, in fact, politics, history, scientific method, astronomy, and biology. This breadth of meaning of literature does not totally avoid the narrowness of the literary criterion, but it makes the disagreement between himself and Huxley appear much narrower. We see the true nature of the disagreement, however, when we recall that for Arnold science, as practiced, was mere instrument-knowledge, like logic and the study of Greek accents. Science, as knowledge, was more interesting, like the details about the likely form of our ancestors. Arnold maintained, however, that even here we were still in the realm of the intellect and no connection could be made between this knowledge and the sense for beauty, or the instinct for right action. For Arnold, literature is a criticism of life and the main instrument of culture; as he says in "Literature and Science," the powers that go to the building up of human life are "the power of conduct, the power of beauty, the power of social life and manners."[37] He was ignorant of the aesthetic impulse and manifestations in science.

Arnold had been unable fully to escape the moral climate of Thomas Arnold or of Victorian England. His formulation of the meaning of culture often emphasizes the moral element, so that despite his call for a return to Hellenizing, he betrays his own concern with Hebraizing.

So the nineteenth-century conflict is laid bare, and we have a prefigurement of the twentieth century's. We still need, however, to give some hint as to what we think was, at this stage, the cultural importance of science. Its cultural impact were many and varied, but we argue that its most pervasive and profound effect was methodological. Huxley himself had a simple view of the structure and function of science, consonant with ours. We shall quote him on this in a later section; however, there is enough of his thoughts on the matter, revealed in this section, to show what importance he placed on method. We might say also that Arnold's failure to appreciate the cultural value of science, as method, is related to his ingrained bookish attitude.

Science and Literature: The Second "War"

We come now to the second major and public conflict between scientists and humanists, which was initiated by Snow in 1959 and escalated by Leavis in 1962. In the light of our thesis about the territorial component of this conflict, it is useful to point to the exalted state of science and scientists in the 1950s. Science was then at the height of its public fame and private enthusiasm. It had ceased being merely an academic subject and was prosecuted in large, sponsored organizations, many of which functioned almost like departments or ministries of government. It had also returned almost to the status of magic, with physicists as high priests of the cult. It was the decade of the DNA (physicists had much to do with this) and of radio astronomy. It was the decade also of rockets and space travel.

It is not necessary to bring out for detailed inspection the theses of Snow and Leavis, for these are well known, having been dissected and warmed over many times during the last two decades and more. We need only to emphasize a couple points made by them.

Snow opened the modern debate through The Rede Lecture, *The Two Cultures and The Scientific Revolution.*[38] Leavis's Richmond lecture, was a response to Snow's. Snow was a man conversant with science and literature. He had been a practicing scientist and was a practicing novelist of contemporary significance. He explored the exercise of power in all its ramifications, in universities and in government. For Snow, science was physics. He was aware of the other sciences but he made only passing references to them; however, he formally widened his terms of reference by describing the *physical* scientists as "the most representative."

Physics is the typical science because the success of its methods has inspired the methodology of the other sciences, and it has invaded the previously neatly demarcated scientific disciplines.

In more general terms Snow, in an article in the *Times Literary Supplement,*[39] sees science as consensual in that past knowledge is so finely incorporated into present knowledge that practitioners do not have to read the original papers of the great ones of the past. This view is similar to Bertrand Russell's, which distinguishes between art and science. He remarked that art is not cumulative, and it therefore needs geniuses; science needs geniuses only at critical moments for the invention of a new world view as we would argue.

The evolution of the practice of consensual science, according to Snow, led to the polarization between the literary and the scientific cultures so that practitioners of the one are unable to reach the first stage of understanding of the other and say "I see." Snow's thesis is that science has become a *separate* culture with the concomitant failure in communication. He implies, nevertheless, that those of the literary culture do not really see science in this "favorable" light, and that they would deny science this autonomous status. Furthermore, they are responsible for the failure in communication. He would probably say, with apologies to Fontenelle, that they despise what they do not know: it is a kind of vengeance.

Snow, however, goes beyond Fontenelle and Huxley and demands a rather special attitude toward science. He says, in his *T.L.S.* article, that "unless we . . . can understand something of science, there will be no hope."[40] Science has become rather encrusted, but it is vital that we should all penetrate to its arcane dogmas and mystifying rituals. We should all share in the culture of science, and be discussing the genetic coding in DNA and the structure of pulsars.

Snow takes this attitude toward science because he believes it has affected the moral lives of practitioners to their advantage and to society's. It has given them compassion and social hope, liberalism and foresight. In his neat phrase, scientists have "the future in their bones," unlike Wordsworth's poet who "looks before and after."

Consonant with the position that science has achieved, Snow exalts it into a new and independent culture, claiming more for it than others had done. One might expect therefore that the reaction of Leavis would be that much more aggressive.

Leavis would be another Arnold. In his confrontation with Snow, he assumes a defensive position similar to Arnold's in the latter's debate with Huxley. He was one of the foremost literary critics in the English-speaking world and propounded with much vigor and conviction that no literary work need be considered that does not possess the moral element to a very high degree.

It is not easy to say what was Leavis's precise concept of science except that science was inadequate for his purposes; we know that he believed that "English Studies" ought to be the center of our cultural activity. By English Studies he does not mean merely literature but something nearer to what Arnold meant by literature. His earlier views made him out to be rather narrow; his literature was Shakespeare and Donne and Baudelaire. When challenged he widened his definition of literature to that of English Studies; he gave us an example the possible study of the changes precipitated in civilization in the late seventeenth century, which would include a history of philosophy, science, and religion and a consideration of French power at the time. Whatever the definition, he maintained that "literature" ought to be the main instrument of culture—that, in fact, the literary culture was the only one. The emphasis was entirely bookish and scholarly. Of course, he could not and did not deny that science has had a cultural impact; he admitted that "science is obviously of great importance to mankind; it's of great cultural importance."[41]

But elsewhere Leavis argued that "the advance of science and technology means a human future of change so rapid" that "we shall need . . . a power . . . of creative response to the new challenges . . . ; something that is alien to either of Snow's cultures."[42] It is a cataclysmic vision that he gives us, and only according to his gospel will salvation be found, that is through literature. Thus Leavis does not even accept Snow's view of what the literary culture is. Indeed, he suggests that Snow's literary intellectual is "the enemy of art and life" and that his culture resides in the despicable, weekly journals and Sunday newspapers.

Leavis had such a jaundiced view of science and such a narrow and dogmatic one of literature that he was probably unsuited to engage in a battle of such seriousness. However, we can use his bad-tempered response as an indication of the serious challenge to literature that science was making from its position of eminence, an indication of a serious territorial dispute.

Both Snow and Leavis were looking at the singularities of the trees and therefore could not see the uniqueness of the forest: Leavis refused even to see some of the trees. Snow, at the same time, was presuming that the fully cultured man should know individual pieces of science rather than arguing that it was the method of science that was culturally important.

Culture and Science

In any case discussion of culture was, for very many people, an embarrassing experience: the word still possessed overtones of a self-conscious and useless activity and also of aristocratic indulgence. However, in recent decades, sociologists and anthropologists have attempted to democratize the concept of culture. The word never had a very precise meaning, which was part of the trouble with it. Recent academic formulations have made it even more semantically amorphous. No one, now, need envy anyone for his culture, or be embarrassed by his own.

The earlier meaning of culture encompassed the intellectual and artistic side of life. The classics of Greece and Rome, music, and painting were the means toward the achievement of this culture. This is a narrow prescription but in the hands of Arnold and Leavis it appears to become even narrower, largely confined as their culture was to books containing the best that has been thought and said. Music and painting seem to play a relatively small part. The latest views, on the other hand, are too wide, and culture's characterization as the whole way of life, too vague.

Cultivation of the mind through books has rightly been widely regarded as too limiting. Cultivation of the affections, for example, must be seen as equally important; if it is worthwhile cultivating anything, it must be worthwhile cultivating the whole person. Music as a pattern of sounds and as resonances of pathos and joy does match literature, whose patterns of words resonate in the same way. Insofar as science is seen as an instrument that helps in the enhancement of some side of our lives, it should also be an instrument of culture. But is it such an instrument?

It is one of the interesting historical points of our thesis that the words culture and science (in their modern sense) and scientist came into use at about the same time in the English-speaking world and, to some extent, in other parts also. This was in the first half of the nineteenth century, and their contemporaneous growth gives some sanction for the continuing interest in the relations between these words or concepts. If words arise to define the existence of things, one can further understand the differences in breadth and force of the debates in the nineteenth and twentieth centuries compared with those of the eighteenth.

We must also note another point of much significance. In the transformation of the meaning of "culture," the emphasis shifted from the cultivation of the soil to that of the mind, and to the pursuit of higher things, generally.[43] At the same time that "culture" came to carry its later import, science (through its developing contact with the practical) began to lose its almost exclusively pure preoccupations and, in fact, ceased to be regarded as (natural) philosophy. "Science" and "culture" were in this sense out of phase.

In this connection Medawar[44] exposed the Anglo-Saxon attitude of reverence for Pure Science as the "by-product of the literary propaganda of the romantic revival," and also as a result of their aesthetic history. Humanists were able to convince many scientists, among others, that their wonderful imaginative activity was akin to that of poetry, which happened to be the "quintessential form of imaginative activity." With this brainwashing, when science and culture went out of phase the humanists could then argue that science's connection was broken—that science was left culturally stranded. We therefore ask the question again, whether science is or can be an instrument of culture.

In any discussion concerning science as culture, we must define the domain of science. Our conception excludes technology. Thus, within this area of discourse, we must distinguish between ideas and methods of science and its products; we must not separate cosmology from telescopes[45] although we shall take account of hydrodynamics while ignoring airplanes. Even if we did not make this distinction and included technology within our definition of science, we would be faced with the historical fact that technology has always been with us whereas the methodology of science is a relatively new phenomenon, only about three hundred years old. Any attempt, therefore, to discover how science thus described has affected our (present) culture would have to explore the effects of this (same) new element, which is just what we call science. Our attempt to investigate Science (rather than technology) seems to be the correct choice, and the distinction we have made both useful and logical.

Within the specified domain we see that there is a general agreement among Arnold, Huxley, Leavis, and Snow that science is a legitimate instrument of culture. The positions of Voltaire and Fontenelle are similar, although Voltaire's is not so clearly marked. They uniformly disagree among themselves, where disagreement arises, in emphasis and according to their professions. The literary men, Voltaire, Arnold, and Leavis, are very concerned about what is happening and what might happen to literature under the assault of science; the scientists, Fontenelle, Huxley, and Snow (who are literary men at the same time), are much more sanguine about the role of science. The (continuing) debate can therefore be seen, in this light, to center around the relative merits of science and literature to function as the major highway to culture; and we can therefore understand why it could have been said earlier that some kind of two-culture concept had been accepted for some time.

The evolution of the dispute can be seen as a classical case of action and reaction, where the increasing claims of the rising scientists have been matched by the increasingly dogmatic assertions of the established *literati* concerning the superiority of literature as an instrument of culture. What is required, it seems, are the recognition and acceptance of the *complementary* nature of both instruments, operating within the same culture. It is relevant, in this context, to talk about the possibility of an ecumenical spirit and the triumph of Fontenelle.

Indeed Fontenelle argues that it was possible to retain all the human qualities while cultivating the sciences. Huxley was, more aggressively, denying that the classicists were "in charge of the ark of culture," and he affirmed that a scientific education was, at least, as good as a classical education for attaining culture. The answer, therefore, is that science can be and has been an instrument of culture—only for some it is not the best instrument.

Even so we must finally ask in what does the deepest cultural influence of this instrument of culture reside? We can better appreciate what it is if we make a further distinction between scientific knowledge and scientific practice, and then compare the cultural impact of both.

One of the major concomitants of the development of science has been the concept of progress. For science has been the activity *par excellence* that has realized this possibility conceived in the recesses of the middle ages. In Darwin's biology, enthusiastically championed by Huxley, we have "biological evolution which eventually provided the theoretical basis for the concept of progressive historical change. The doctrine of evolution therefore provides one of the most striking examples of the influence of scientific knowledge on modern culture."[46] So we have an example of how scientific knowledge has had a cultural impact.

If we turn to Huxley's grandson, Aldous, we find him relating how ethologists disabused poets, from many ages, about the source and the meaning of the nightingales' song. The song does not have for nightingales all those romantic messages beloved of poets, but only a practical and "unpoetic" one. Huxley remarks that the ethologists being "able to recognize this truth and to act upon it represents a major triumph."[47] Here we have examples of the impact of scientific knowledge on major areas of our lives, and Snow would have our (genuinely) cultured person in possession of such knowledge. The facts are different, and, sadly scientific *knowledge* does not play such an integral part in the lives of most (educated) people; it is negligible in the lives of the so-called average man.

With scientific practice the situation is different, and although Arnold is accommodating in admitting scientific method into his cultural canon, we have to go to Huxley to find the canonical argument. Huxley had a simple view of the structure and function of Science. For him "Science is not . . . a modification of the black art, suited to the tastes of the nineteenth century, and flourishing mainly in consequence of the decay of the Inquisition. Science is . . . nothing but *trained and organised common sense*. . . . So, the vast results obtained by Science are won by no mystical faculties, by no mental processes, other than those

which are practised by everyone of us. . . . The man of Science . . .
simply has with scrupulous exactness the methods which we all
. . . use carelessly: and the man of business as much avails
himself of the scientific method . . . as the veriest bookworm of
us all."[48]

In this passage Huxley puts his finger on an essential nature, a
distinguishing feature of science; and we would maintain that to
the extent that it is scientific *method* (unlike scientific *knowledge*)
that has permeated our conscious and subconscious activities, to that extent indeed has science had its most profound
and lasting cultural impact on us *all*. We shall elaborate on the
point.

We do not believe that it is possible to make *consistent* claims
for any profound cultural impact of scientific knowledge, for such
knowledge has failed to command the everyday imaginative attention even of scientists themselves. At the same time, scientific
method has done so.

The inconsistency we hinted at arises because the sense of the
theories of Darwin and Freud, for example—that is, their scientific knowledge—has permeated even modest intellectual strata
of Western society. We look at our relatives, friends, and ourselves
with the eyes of Freud and Adler. But it must be unusual even for
a scientist, indeed, to look at our world with a steady Einsteinian
gaze. Furthermore, even within the physical sciences distinctions can be made. There are those theories that bear down on us
even in our privacy so as to modify our understanding of our
position in the universe. Such a theory is cosmology. Other theories, such as that of superconductivity, even scientists ignore
except in their strictly professional role. There is no consistent
impact of scientific knowledge, and therefore our culture is
rather thinly leavened with what I have called scientific knowledge.

What we may consistently maintain, however, is that the
method (rather than the substance) of science has entered into
our culture. Sypher makes the point about the history written by
La Popelinière in the seventeenth century. Plumb, the historian,
assures us that science has made, through its methods, monu-
mental contributions to history.[49] It is true also for literary criticism, economic analysis, philology, commerce, and even sport.
Scientific method is an integral part of our culture.

Summary and Conclusion

The two-culture debate has been taken back to the time of
Fontenelle and Voltaire in the early eighteenth century, not be-

cause there was conflict and debate at that time but because the possibility of such a conflict was clearly seen and articulated. This makes us aware of the origins of the two major debates of the nineteenth and twentieth centuries. When we recall that science as an autonomous (and successful) enterprise was being established in Fontenelle's time, and that it would enter a field then dominated by literature, then the inevitability of the conflict is historically and profoundly understood. The historical account has been the inspiration for this chapter.

It has been implicit in the debates that science has or can have a cultural impact; and this has been achieved despite the multiplicity of cultures, because of the pervasiveness of science. This is undeniable, and its creative impact upon literature itself right into the twentieth century has been well documented, from Milton to Proust, from Donne to Virginia Woolf. However, we maintain that this impact has not been pervasive enough. In other areas of life, such as entertainment and recreation, the impact might be even more sustained and pervasive, but the source of this impact is technological, which we ignore for already stated reasons. Through scientific method only can we see science as an integral part of a whole culture. Of course, once we accept the larger perspective entailed by this argument, the rivalries of the nineteenth and twentieth centuries appear petty, futile, and even unnecessary. Yet we come finally, from the same historical narrative, to the realization that the debates, futile though they were, were historically inevitable and therefore comprehensible.

Our narrative account of the developing debate over the centuries suggests a petty, not to say, base motive concerned with territorial advantage, and to the extent that this was the case the debate was insubstantial and ignored the issue of the real nature of the cultural impact of science, both actual and potential. However, even if allowance could be made for motive, the nineteenth-century debates still had an air of futility, for it was not as if there was a choice as to whether science would be pursued privately as it had been in previous centuries, or made public and therefore become public property. *Publication* had become one of the essential (sociological) features of the pursuit of science. Even for those burrowing away in their lonely studies, their work achieved its rationale only in its *publication*. The cultural dynamics of such an activity was unstoppable.

It is our further point that an analysis of the structure of science into Science and Technology, and a further refinement of

Science into scientific knowledge and scientific practice, as we have done, would have made the two-culture brawl of the 1960s unnecessary. The recognition that not even scientists are culturally affected by their scientific knowledge would have prevented Snow from making his exaggerated requirement on everyone to know the second law of thermodynamics.[50] At the same time he would have recognized that he was barking up the wrong tree; there was no uncivilized prey lurking in its branches: we had all been culturally grounded within the scientific practice or method of centuries. This was science's cultural legacy and we had all received our share.

6

Time and Reality in Eliot and Einstein

> Time present and time past
> Are both perhaps present in time future,
> And time future contained in time past
> If all time is eternally present.
>
> —T. S. Eliot, "Burnt Norton"

> It is a world like Lobatchevsky's; the worlds created by artists
> like Johnson are like systems of non-Euclidean geometry.
> They are not fancy, because they have a logic of their own; . . .
> and this logic illuminates the actual world, because it gives us
> a new point of view from which to inspect it.
>
> —T. S. Eliot, *The Sacred Wood*

Introduction

In the previous chapter we examined, among other things, the general impact of science on culture as part of the (larger) section's attempt to argue for the cultural nature of science. The decisive element we identified was methodology, but it was obvious from the narrative that cosmology, for example, and the biological sciences also weave perceptual and emotional changes of lasting value into our cultural lives. This does not mean that science, in whatever mode, always affects the wider society. Indeed, we hinted at such a denial in the previous chapter. Thus we have, for example, the general modes of space and time as currently understood through Einstein's theories, influencing in a particular way the creation of much art, most of which we all would agree has had no general impact. Yet, we wish to propose that this very aspect of science has left us its legacy. We can support this proposition by distinguishing between the creative and the journalistic, between the lasting and the ephemeral, and by showing that it was not just a literary fashion that was being created in Eliot's work, for example. In this chapter, therefore, we

130

shall be concerned with the particular rather than with the general cultural impact of science, as the title of the essay reveals, and we shall seek it in the work of T. S. Eliot as influenced by Einstein's theories of relativity.

We have always been fascinated by the concepts of space and time; and a gauge of this fascination is the volume of literature they have provoked. It would be true to say, however, that *time* has been the more intriguing; it seems to be more fully woven into our personal lives. Time is associated with "the permanently lost," "the inevitable," and "the inaccessible," and philosophers from Aristotle to Zeno and their heirs have attempted either to make compliant this *insouciant* maiden, or to escape the bondage of their infatuation. Probably only the mystics have succeeded.

This interest has naturally found its way over the centuries into the core of much fiction; the present century bears testimony to such an invasion, in the work of such writers as Proust, Joyce, Yeats, Wells, and Mann. From the writings of Bergson to the climax of the publications of Einstein's theory of general relativity in 1916, there was a reinvigorated interest in the "mystery" of time, and since those decades there may have been shifts, but no abatement, of interest in the subject. It was a revolutionary account of the nature of time that emerged from Einstein's theories, an account that simply could not be ignored by the philosophers.

This being so, it seems appropriate to investigate the influence or, at least, the concordance of thought between the greatest revolutionary scientist of the present century and the greatest revolutionary poet of the same era. This cross-fertilization does not undermine, but can be set against, the pessimistic account, as we saw in the previous chapter, that Snow has given us of the cultural-scientific impoverishment of nonscientists. Many have noted the influence on men of letters of scientific advances and discoveries at other times. Marjorie Nicolson, to whom we made reference in a previous chapter, has written at length of the influence on Donne, Milton, and Swift of the "new science" of the seventeenth century. The present, less ambitious, essay gains its justification within the rationale of the chapter, because here a particular scientific concept espoused by a particular man is seen to have *permeated* the work of another. Furthermore, to the extent that Eliot's poetry is accepted among the classics of our heritage, so has this scientific concept of time influenced our culture. Although "time" ultimately uncovers so little of its mystery to us, it is probably only through the concentration of great poetry that the essence of such a profound concept can be more widely appreciated.

Many explications of Eliot's poetry have drawn attention to his concern with time. In our investigation of his poetry and other writings—which is intended to discover his views on the nature of time and reality—these explications provide a useful reference. First, we shall attempt to establish Eliot's acquaintance with some of the necessary concepts of geometry and relativity before showing in what ways the challenging implications of Einstein's theories are consonant with or affect his own views. Then, Eliot's philosophical attitudes, especially those to be found in *Knowledge and Experience,* will be discussed in the light of his poetical statements on the status and meaning of time and reality. This exercise will establish a certain coherence and consistency in Eliot's thought over a long period of time, and at the same time discourage any views that his concern was *merely* with the topical, journalistic aspect of Einstein's theories.

The Nature of Time

Discussions on the nature of time have often centred around its substantiality—that is, whether time exists as an independent thing, as a specific temporal entity, or merely as an attribute of, or a relation between things or events. These aspects of time have been displayed in the typical discourse on the absolute versus the relational nature of time. Discussions have equally often centered around the objective (physical) nature of time versus its subjective (mental) nature. Within these frameworks scholars have also argued in terms of the categories of appearance and reality (of time).

Additional subsidiary features emerge, revealing the pronounced psychological component of time, in which the flow varies according to bodily and mental states. As a result, mortals have certainly wished that they could arrest the flow of time, revoke their past, and gain a window into their future. We are not concerned with this issue.

What will particularly concern us are the age-old contraries of "Being" and "Becoming." Time is seen by some as partaking of the "static temporal," and by others of the "dynamic temporal." In essence, events are regarded as being either simultaneously given in time in a (serial) static order, for which the relevant relationship is "before" or "after," or as emerging in time and exhibiting a flux as they go from the past through the present into the future. In the latter view events come into being, or "become." The subsumption of the latter of these two philosophies is that time is a series of instants, like points on a line, stretching from $-t$ to $+t$, the "now" being one point separating past from future. This model raises certain awkward questions, because the real instants cannot be points: events take time and do not occupy a zero-extensive temporal region. The theory of relativity has added fuel to the fires of discourse, for it can be interpreted as regarding events as *existing* in a four-dimensional space-time continuum. Superficially, at least, it gives support to the philosophy of Being.

The issues introduced above have been debated from the time of Aristotle, who argued for the insubstantiality of time, and of Zeno, who espoused a philosophy of Being. Newton adopted an objective and absolute view of time, whereas Kant, who also believed in its absolute nature, denied the objectivity of temporal becoming. For him, furthermore, the concept of time was given a priori, the condition for, and not the result of experience. Bradley did not ascribe any reality to time, whereas Bergson subscribed to the reality of a psychological time, what he called "duration." All the relevant arguments may be regarded, as Čapek remarks,[1] as echoes of the original Parmenidean-Heraclitean dispute, in which same tradition we may see the more recent division between philosophers like Harris, William, Weyl, Grünbaum, and (probably) Einstein on the one hand, and Bridgman, Eddington, Čapek, and Sellars on the other.

Einstein's Theories

Despite the unending nature of the debates, it is instructive to document the ways in which a great thinker like Einstein and a fundamental theory like relativity have influenced the discourse in our time. The authority of Einstein's conclusions is seen to full effect in the way both sides of the debates have used these conclusions for the support of their own case. This is especially true of that aspect of the theory of relativity that establishes the relativity or simultaneity, and represents the world as a four-dimensional space-time manifold. Both sides (of the Being-Be-

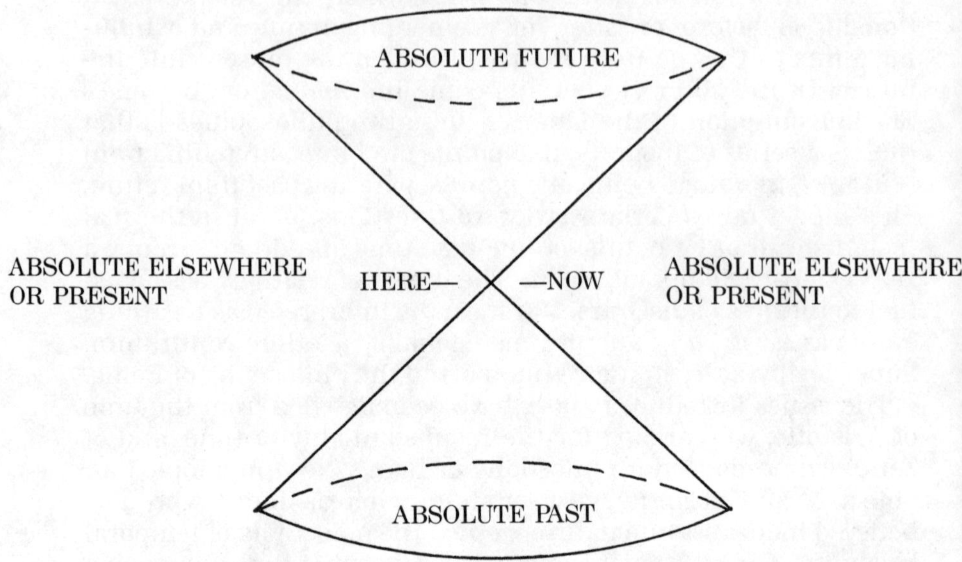

coming debate) have appealed to what have become the higher courts of science.

The relativity of simultaneity concerns itself with separated events and implies that the measured time between two such events may be different for two observers in relative motion. In particular, two events simultaneous for the one observer will not be simultaneous for the other. Such observers will certainly disagree among themselves as to what is the present, the "now," or what we are accustomed to think of as the point of intersection of the past and the future. Equally significant is the fact that for these observers the comprehensive and absolute order "before and after" is destroyed. The absolute nature of time dissolves in this "chaos." A new picture of the structure of time emerges, of which Eddington and others have given the representation shown in the figure.[2]

We no longer have a straight-line representation of past, present, and future. The relativity of simultaneity posits a disagreement between observers in relative motion concerning the "now," and the separation of absolute future from absolute past is a measure of this disagreement, which depends on the separation of the events. Within the wedge-shaped regions illustrated, there are no events that can influence us or that we can influence. Thus these regions satisfy our usual concept of the present. This concept embraces the proposition that nothing I do at this moment can affect my looking at the paper in front of me (now). Any such act can have its effect only at some different instant.

Of course, events are space-time points, so that we designate the intersection in the diagram as "here-now." Observers dispute about the "now"; they do not disagree, however, about the "here-now." There is no disagreement about events that are not only simultaneous but also coincident, and no two observers would deny the simultaneity of the events of X's death, the piercing of his heart, say, and his backward stumble. "Here-now" is always and everywhere; it is absolute. We have thus made a distinction between "here-now" and "now." The latter is no longer a simple dividing line in space but the whole wedge-shaped "elsewhere"; the former is the reality, and the world is the manifold of all such events.

The space-time ontology was certainly encouraged by the Lorentz transformation equations of Einstein's special theory, although the extreme interpretation of an indissoluble mixture may be readily countered.[3] The milder view was further enhanced by Minkowski's formulation and by Einstein's adoption of this geometrical formulation for his general theory. Time, as a formal consequence, was spatialized beyond our wildest verbalizing. Now Lobachevsky, Bolyai, and Riemann had shown more than fifty years before Einstein's theory that non-Euclidean geometries were logically perfectly possible; Einstein's theory of general relativity, in fact, presents the world as a four-dimensional matrix of objective space-time events obeying a particular form of Riemann's generalized non-Euclidean geometry. The formation of this theory undercut the traditional view of the physical world as necessarily obeying the laws of Euclidean geometry.

The above support for the view that the world is (existentially) made up of space-time points has led some philosophers to the interpretation that things do not come into the world of space. According to them,[4] our death, as an event, already exists in the world of space-time; it is part of the large canvas of existence that we are unable to "see" because of our limited view. Other commentators have rejected this "absurd" philosophy of Being. Čapek, for example, maintains that because observers of separated events will disagree about the simultaneity of events—that is, about the cosmic "Now"—there is therefore no such thing as "Nature at an instant." Indeed, he argues that the theory of relativity supports the philosophy of Becoming, and regards the more obvious separation of past and future in Eddington's representation of time as the pictorial counterpart of this philosophy. The region "Elsewhere" represents a more vivid emergence from the past into the future and, Čapek believes, so does the established phenomenon of the expansion of the universe.[5]

We are not, here, particularly concerned with which interpretation is correct, but certainly with the way in which Eliot has made use of the new ideas and himself interpreted them.

Time in Eliot's Early Poetry

Having filled in the background of our picture, we shall now sketch a couple of important details in the foreground. The first includes a selective documentation of Eliot's interest in various aspects of time; the second establishes his acquaintance with Einstein's theories of relativity.

Eliot's concern with time was pervasive and profound. A good case can be made, when we realize that this concern was not confined to his mature poetry nor even to what may be called his definable career. Among the poems written in early youth we find "A Lyric" and "Song."[6]

> If Time and Space, as Sages say,
> Are things which cannot be,
> The sun which does not feel decay
> No greater is than we.
> So why, Love, should we ever pray
> To live a century?
> The butterfly that lives a day
> Has lived eternity.
>
> If space and time, as sages say,
> Are things that cannot be,
> The fly that lives a single day
> Has lived as long as we.
> But let us live while yet we may,
> While love and life are free.
> For time is time, and runs away,
> Though sages disagree.

Thus from the very beginning Eliot was attracted by the problems of space and time and the relationship between time and eternity.

He was, furthermore, aware of the philosophical arguments that surrounded these subjects, but emotionally learned toward the unsophisticated view

> For time is time, and runs away,
> Though sages disagree.

His creative work from *Prufrock* in 1917 to "Little Gidding" in 1942 provide mounting evidence for this conclusion. His sensibilities were aroused by the flavor of a particular time or the passage of the seasons, as in the "Preludes":

> The winter evening settles down
> With smell of steaks in passageways
> Six o'clock
> The burnt-out ends of smoky days.
>
> The morning comes to consciousness
> Of faint stale smells of beer.

There is here the specificity that comes from sensitive observation of particular times. In *Portrait of a Lady*, "December afternoon" precedes "April sunsets," "August afternoon," "October night." The sequence of the seasons is evident also in the first passage of *The Waste Land,* and the *Four Quartets* can be placed almost in a one-to-one relationship with the four seasons. In *Prufrock* we find his more famous correlatives, his other "withered stumps of time."

> I have measured out my life in coffee spoons.
> . . . .
> I grow old . . . I grow old . . .
> I shall wear the bottoms of my trousers rolled.

There is more than mere reference in those lines. In the passage following the extended "yellow fog" metaphor in *Prufrock,* Eliot also captures the infinitude of (psychological) time through the expansion of the universal moment of indecision to encompass "all the works and days of hands." Some of his effects are brilliantly achieved by characteristic use of repetition. In this moment of indecision

> There will be time, there will be time
> To prepare a face to meet the faces that you meet;

Furthermore, "The Rock" gives us at a certain point a complex play on the word "time" that yields, simultaneously, an analysis of the Incarnation.

> Then came at a predetermined moment, a moment in time and out of time,
> A moment not out of time, but in time, in what we call history: transecting, bisecting the world of time, a moment in time but not like a moment in time,
> A moment in time but time was made through that moment: for without the meaning there is no time, and that moment of time gave the meaning.[7]

Here we have punctual time within a progression, time as history, and time set within the objective-subjective duality. Time that "we call history" looks forward to "Little Gidding," where Eliot describes history as "a pattern of timeless moments."

The Quartets themselves may be said to be about time, without in any way detracting from their rich complexity. In them, Eliot explores the nature of the time of immediate experience, the time of the turning of "old stone to new building, old timbers to new fires," and the time of eternity. Eliot was clearly haunted by time, and in the *Four Quartets* he attempted to articulate the means for transcending it. He asserts in "Burnt Norton":

> Only through time time is conquered.

It might not be true to say that only a philosopher could understand the poem, but it is certainly true that Eliot's poetic treatment of time is fundamentally philosophical in its exploration of the grounds of our perception of time and in its analysis of time's manifold manifestations.

Einstein and Eliot

We shall now attempt to complete the second task that we set ourselves in the previous section: to establish the degree of awareness that Eliot had of Einstein's work.

The second quotation at the head of this chapter indicates that Eliot was at least acquainted with the possibility of non-Euclidean geometry and, indirectly, with the nature of space and time that emerged from Einstein's theories. As we have seen, one of the essential modifications made by Einstein's general theory is that the physical universe is fundamentally non-Euclidean.

It is very likely that Eliot was not deeply versed in general relativity. Who was, in those days? Indeed, a closer examination of the essay on Johnson reveals that there is less force in Eliot's reference to a world like Lobachevsky's rather than to one like Riemann's. Both are non-Euclidean, but Einstein found that our world was like Riemann's more general formulation. This included the particular Riemannian world of constant positive curvature as well as Lobachevsky's world of constant negative curvature. It is possible that Eliot's choice of reference indicates a less than full appreciation of the finer points.

However, inspection of this essay also reveals that Eliot was not just name-dropping; Eliot is discussing the verse of Beaumont and Fletcher, which he thought to be

> superficial with a vacuum behind it; the superficies of Johnson is solid. It is what it is; it does not pretend to be another thing . . . ; a man cannot be accused of dealing superficially with the world which he himself has created; the superficies *is* the world. Johnson's characters conform to the logic of the emotions of their world. It is a world like Lobatchevsky's; the worlds created by artists like Johnson are like systems of non-Euclidean geometry. They are not fancy, because they have a logic of their own; . . . and this logic illuminates the actual world, because it gives us a new point of view from which to inspect it.[8]

The passage shows Eliot's awareness of some of the technical reasons for the possible existence of non-Euclidean geometries—namely, that one could modify a postulate as one wished, and if the *logic* of the exercise were sufficiently precise and powerful, a perfectly possible non-Euclidean geometry would be established. As Eliot maintained, once Johnson's characters conformed to the *logic* of the emotions of their world, they were fully adequate, nonsuperficial characters. It is in this correspondence that Eliot's reference to non-Euclidean geometry is to the point, and quite sophisticated.

Furthermore the quotation below reveals a philosophy of space and time that does not subscribe to their indissoluble mixture, as some would argue on the basis of Minkowski's four-dimensional formulation of Einstein's special relativity theory. In the emphatic statement of this youthful poem

> . . . I know that time is always time

> And place is always and only place[9]

Eliot implies that he is aware of some of the philsophical conclusions or interpretations of Einstein's theory, but maintains
the commonsensical and traditional view consistent, perhaps,
with his undeveloped philosophical position.

Time and Reality in Eliot's "Four Quartets"

We are now in a position to give some flesh and blood to the
bare bones of our title. There are two levels at which we shall
attempt this exercise. The first will display occasions on which
Eliot makes use of the vocabulary of the theories of relativity. The
second will show us instances of significant, connected use of the
idioms and grammar of these theories. Significance will be seen
to reside mainly in the possibility of establishing, out of these
instances, a philosophy of time and reality that is Einsteinian.
Mere coincidence of views, of course, is no proof of influence, but
the context of the coincidence does imply some influence.

Eliot's poetry is rich and subtle in its allusions, and Blamires
has made, at length, the point that the *Four Quartets* is about
echoes. (This poem will be our main source in this section, and
we shall find many echoes indeed.) In "Burnt Norton," Eliot
introduces a rather fractured view of a wonderful experience in
the rose garden. To capture the priceless moment we are en-
joined:

> Quick, now, here, now, always

with the ensuing lament,

> Ridiculous the waste sad time
> Stretching before and after.

Experiences like this (which alone give meaning and constitute
reality) are described by Eliot in "Little Gidding" as

> . . . the intersection of the timeless moment
> . . . England and nowhere. Never and always.

When, in "East Coker," Eliot contemplates at the sea's edge, at dawn, the passage of his ancestors in time and place, he sums up his situation cryptically.

> . . . I am here
> Or there, or elsewhere. In my beginning.

We are struck in these passages by the echoes of the Eddingtonian construction of Einstein's world, categorized into "Here," "Now," "Here-now," and "Elsewhere." These categories clearly have commonplace associations, but their juxtapositions strengthen the sense of the direction of their echoes. Eliot uses the techniques of echoing and allusion so extensively that we may be allowed to suggest that it is done with deliberation. We may also be allowed to make the point that, according to the quotations from "Burnt Norton" and "Little Gidding," "Here-now" is everywhere and always, which is the conclusion arrived at in our earlier analysis of the Eddingtonian construction. Indeed, when we dip below the surface, we find that Eliot does subscribe to a relativistic view of time and reality.

At the very beginning of "Burnt Norton," he speculates that

> Time present and time past
> Are both perhaps present in time future,
> And time future contained in time past
> If all time is eternally present . . .

At the end of the movement he puts forward, less speculatively that

> Time past and time future
> . . . .
> Point to one end, which is always present.

Eliot is thus leaning towards a static view of time.

Now, analysis of the structure of the individual quartets and of the *Quartets* as a whole brings out the fact that final statements represent syntheses of Eliot's views. At the very end of "Little Gidding," for example, he talks sanguinely about the time

> When the tongues of flame are infolded
> Into the crowned knot of fire
> And the fire and the rose are one.

Within this context, one apprehends Eliot's view of time, wherein reality is a copresence of all time—past, present and future. Later on we shall see that his general philosophical views are unfettered by any vestiges of a separate reality for the various categories. The copresence of all time can clearly be interpreted as spatial, in the simplest description, because the locations of space are thought of as existing simultaneously. In the last movement of the "Burnt Norton" quartet, Eliot writes:

> Or say that the end precedes the beginning,
> And the end and the beginning were always there
> Before the beginning and after the end.
> And all is always now. . . .

Two important comments can be made on these lines. The first is that they take us into realm of the Eddingtonian "Absolute Elsewhere or Present," where "before" and "after" lose their absolute relationship. The second is that the phrase "all is always now" takes us into the realm of Being. The philosophy embedded in this line may be taken as Eliot's final statement about the copresence of all time.

It is possible to penetrate deeper into Eliot's world and arrive at the units of reality as he sees them; only thus can we complete the comparison with Einstein's world. The three quotations in the second paragraph of this section indicate a space-time view of reality—a view achieved not merely through the juxtaposition of words connoting space and time but, more significantly, through the integration implicit in the identification of the intersections (which are reality) as both place (here, England) and as time (never, always).

We remarked before that for Eliot reality is apprehended through the consciousness of those (timeless) moments in the rose-garden. One aspect of reality, history, "is a pattern of timeless moments." Furthermore, reality is all and always about us for

> Here the intersection of the timeless moment
> Is . . . always.

Yet we achieve but a limited view of reality for

> . . . to apprehend
> The point of intersection of the timeless
> With time, is an occupation for the saint—:

Nor is this our only failure; the pattern of the quintessential moments in the rose-garden is often denied us for

> . . . human kind
> Cannot bear very much reality.

To see more clearly how Eliot's vision meshes with the conclusions of relativity theory, we should note how the two images of reality are merged in "Burnt Norton."

> . . . there the dance is,
> But neither arrest nor movement. And
> do not call it fixity.
> Where past and future are gathered.

The axle-tree as a representation of reality is no mere limited spatial or temporal metaphor: it embraces space and time, and also symbolizes the intersection of the timeless with time. When Eliot goes on to say

> . . . Except for the point, the still point,
> There would be no dance, and there is only the dance,

we conclude that the space-time units and the pattern they form *are* the only reality. These units of reality, these events like "the moment in the rose-garden" or "the moment in the draughty church at smokefall," form a pattern in the same way that Einstein's units of reality, his (space-time) events, form the matrix of his four-dimensional manifold.

Thus there appears to be a *prima facie* case for regarding Eliot's views of space and time as identical with those expounded at the same period by various Einsteinian relativists. In particular, there is evidence both for a subscription to the relativity of simultaneity and to a "static temporal."

Eliot and Others

Whether there is a *prima facie* case or not we must be seen to be relatively unbiased in trying to establish the particular connection between Eliot and Einstein: it might be that Einstein's influence on Eliot is merely a minor episode in his work and that others have had a more profound influence and effect. Indeed,

Einsteinian resonances are not the only ones to be found in Eliot's poetry. Other philosophical concepts and arguments are represented in his poetry and plays, and they extend our knowledge of his philosophical inquiry. He was critically concerned with the work of certain philosophers, especially with St. Augustine, Bergson, and Bradley, but we shall show that their theory of time had not much consonance with Eliot's own views.

In "Burbank," and in the Fire Sermon of *The Waste Land*, it is easy to identify Eliot's acquaintance with St. Augustine. St. Augustine's known difficulties with the concept of time arose from his attempt to account for our measurement of something that exists only in a durationless present. This led him to reinforce his subjective view of time by giving it a psychological bias. Like Aristotle, St. Augustine saw time as an aspect of things rather than as some independent reality, but whereas Aristotle regarded time as inextricably embedded in motion, St. Augustine counterproposed that stillness was also measured in time. This led to certain paradoxes, and caused the desperation of his cry, "What then is time?" In his concept of time, there was no *reality* to past, present and future; as he said, "there be . . . a present of things past, a present of things present, and a present of things future."[10] This philosophy is consistent with his view of a punctal present that is only an ideal (mathematical) point. He arrived at this view from an analysis of whether a period of one hundred years could be "present," from which he concluded that "neither things to come or past are." Only the present is, and "The present hath no space."[11]

Within this framework, St. Augustine adopted a philosophy of Becoming. This is not in conflict with his copresent reality, which is existential in quality and does not deny that the past was and that the future will be. Indeed, when he compares the times of man and of God he asserts that the years of the "future, . . . when they come, they shall be past; our . . . years both come and go, that they may all come." We are clearly in a state of flux. What is static is the time of God, or Eternity, which is a coeternal present. For St. Augustine, "the present, should it always be present, and never pass into the past, verily it should not be time, but eternity."[12]

Reality for St. Augustine resides, therefore, in the punctal present, which is elusive, unlike God's present which is coeternal. However, it is possible to have a meaningful experience of such a moment (which may be seen as an atom of eternity) by seeking out the still center of one's being. There, at the intersection of the temporal and the timeless, the still point, St. Augustine himself experienced aspects of eternity. Words and images that appear in St. Augustine's philosophy certainly echo in Eliot's work, but Eliot expands the concept of copresence beyond the punctal present of St. Augustine. There is also a common concern for the relationship between time and eternity, and a ready correspondence between St. Augustine's still center and Eliot's axle-tree. However, in the final analysis we must declare their difference, for while St. Augustine subscribes to a philosophy of Becoming, Eliot's views can more easily be fitted into a block universe supported by a philosophy of Being.

When we closely examine Bergson's philosophy of time, we find less sympathy with Eliot's. The central aspect of Bergson's philosophy is that time must be perceived as duration, which he interpreted as entirely mental. Furthermore, Bergson strongly denied the homogeneity of time, allowing this property to space only. This denial is consistent with his dismissal of outer or clock time as unreal; for him the true time was duration, or inner time, which could therefore not be spatialized or homogenized. This is consistent with his philosophy of Becoming. As he wrote in *Time and Free Will,* "when we make time a homogeneous medium in which conscious states unfold themselves we take it to be given all at once."[13] But for Bergson, consciousness was the note of the present, the present moment being constitued "in that continuity of becoming which is reality itself."[14] He thus rejected a block universe, implied by temporal homogeneity, and "in which past, present, and future coexist simultaneously," because he could not see how a homogeneous (spatialized) time could account for temporal succession and a philosophy of Becoming.

Bergson's *Creative Evolution* argues extensively for such a philosophy of Becoming: "in reality there is no form . . . what is real is the continual *change* of form."[15] Eliot is known not to have set much store by evolution and its implications of progress, and in "The Dry Salvages" he makes a knowing and weary dismissal of its claims, implying, at the same time, his own philosophy of Being:

> It seems, as one becomes older,
> That the past has another pattern, and ceases to be
>   a mere sequence—
> Or even development; the latter a partial fallacy
> Encouraged by superficial notions of evolution.

Thus the internal evidence does not suggest a convergence of views between Bergson and Eliot. But Eliot was even more specific; he thought that "the Bergson time doctrine . . . reaches the point of fatalism which is wholly destructive."[16]

It is one thing for Eliot's philosophy of time and reality to emerge almost entirely from the shadows of St. Augustine and Bergson. It is another thing to see it fully in the Einsteinian light. It is therefore helpful that Einstein's informal pronouncements illuminate Eliot's poetical statements, however indirectly. Einstein has been quoted as saying that "becoming" in the three-dimensional space has been transformed into "being" in the world of four dimensions. Furthermore, as Čapek has observed, Einstein's leaning towards Gödel's idea of the observer visiting the past and returning to the present suggests a predilection for the static view and a philosophy of Being.[17]

The interpretation of Eliot's philosophy of time and reality and its Einsteinian connection are supported at the fundamental level of explicit philosophical statement in *Knowledge and Experience*. In this work, Eliot does not take up a dogmatic position on the nature of reality, nor does he hold in view any essential dualism, as Kant did between nature and spirit or subject and object. He was to a large extent in the tradition of idealist metaphysics, and like Hegel, he adopted the dialectic method.[18] In discussing "truth", for example, Eliot asserted that "the true critic is a scrupulous avoider of formulae . . . he finds fact nowhere and approximation always."[19] Thus any experience can only be fragmentary and interpretive. In fact, he believed that "every interpretation, along perhaps with some utterly contradictory interpretation, has to be taken up by every thinking mind."[20] If one were to ask what room there might be, on the basis of these statements, for a philosophy of the absolute, Eliot's retort would be that one was unaware of the "dialectic process," a process that was at work in the whole of the *Four Quartets*.

This is the method he uses in his approach to reality through an analysis of knowledge and experience. No polarity is established between (knowledge of the) object and (experience by the) subject, for "an object of knowledge exists only in a context of experience, . . . with which it is continuous."[21] Thus *knowledge plus experience is the integrated reality.* In the same way, reality is neither real nor ideal: these two terms simply represent different expressions of the same thing.

Eliot's undogmatic view of the nature of reality is emphasized by his statement that "From first to last reality is experience, but experience would not . . . be possible without attention and the moment of objectivity";[22] the objects of knowledge, however, do not gain their objecthood from their qualities. The object is simply the "point of *attention:* that which an electron, a Balkan league, my table, whiteness, have in common as objects is just the *moment of objectivity.*"[23] These expressions also focus on his linguistic usage and his meaning, in the *Four Quartets,* where his points of attention are transformed into those intense moments of consciousness, the pattern of which is reality. This establishes one link of continuity between his philosophical statements and the implications of his poetry. Another is established by his view that the only independent reality is immediate experience, which he calls "a timeless unity which is not as such present either any-*where* or to any*one.*"[24] It is that atom of eternity which is experienced at

> . . . the intersection of the timeless moment
> . . . England and nowhere. Never and always.

In *Knowledge and Experience,* Eliot pursues his concern with time and eternity more deeply, especially when he attacks the problem of time in knowledge and in reality. In particular, he poses a dilemma: existent things gain our attention, which is a temporal process, and therefore such objects are in time; yet the reality of the object is its holding together in various moments of attention through *meaning,* which is a subsistent thing and therefore not in time. How Eliot resolves this dilemma is not of immediate concern to us. What is interesting is the similarity between this formulation of a network of moments of attention as providing meaning, and that of the *Four Quartets,* where only the pattern of those intense moments of consciousness possesses any meaning.

Conclusion

In his poetry, Eliot extensively used both the vocabulary and the grammar of relativity theory as expounded by Einstein, Eddington, and others. We have tried to show that, from his adolescent to his mature poetry, Eliot indicate an abiding concern with the nature of time. Thus, his interest in the philosophical implications of Einstein's theories can almost be taken for granted; it extended even to such a fringe work as Dunne's *An Experiment with Time,*[25] which relies heavily on the implications of

Einstein's theories. Nevertheless, in the preface to *Knowledge and Experience,* Eliot disclosed that "Forty-six years after my academic philosophizing came to an end, I find myself unable to think in the terminology of this essay. Indeed, I do not pretend to understand it."[26] He has also said that a poet might use philosophical ideas, but he has no need to think them. I would agree with Wollheim[27] that this is a piece of ironical self-deprecation; in view of Eliot's intellectual history, we can certainly conclude that he thought as well as used philosophical ideas and was therefore susceptible to the influence of Einstein's theories (negatively or positively)—or to any others that concerned his abiding interest in the nature of time and reality.

Eliot's view on reality seem to have drawn some inspiration from Bradley. However, *Knowledge and Experience* reveals views that we may take to be his own, only tempered and made articulate by his reading of other philosophers. He has fitted into a tradition that George[28] has called existential, but nevertheless, Eliot's firmly stated Bradleyan positions place him in the monistic tradition that goes all the way back to Parmendes's philosophy of Being.

The foregoing account permits us to make a more balanced appraisal of the continuity of Eliot's thought, its traditional limits, and its modern postures. In particular, the philosophy of time entailed by his poetry seems effectively to spatialize time in the way that relativity theory does; it encompasses a conditional present that is the natural result of the relativity of simultaneity. His *Knowledge and Experience* has given us, furthermore, some indication that his views of the nature of science coincided with those of Einstein. Eliot believed that natural science's task was not to decide the relation of appearances to an ultimate external reality. What it does, he says, is to present us with a large canvas of appearance that is a totally valid aspect of reality, only referred to the subjective side of experience. This datum is a system of terms in relation, and the terms are defined by their position in the system; they have no characteristics that could be isolated from the system. We need only draw attention to the fact that Eliot's description is perfectly valid for the space-time manifold of events in special and general relativity, in which relations define the reality.

General relativity stops short of asking what is the ultimate reality behind the non-Euclidean manifold; it is, of course, concerned with ontological questions of a different order. We argue, with Fraser[29] that "The ontological enquiry of Relativity Theory may be summed up in the questions, "What is then?" and "Where is there?" It is clear that Eliot subscribes to this Einsteinian view when he says, in "Burnt Norton":

> . . . and there is only the dance.
> I can only say, *there* we have been: but I cannot say where.
> And I cannot say, how long, for that is to place it in time.

Eliot is obviously asking the same ontological questions, for an answer to the implied temporal question "How long?" depends on our being able to answer the other question, "When is then?" Thus it brought together Eliot's philosophical and scientific views, and his poetical practice also, in one substantial Einsteinian embrace.[30]

In an earlier section we made the point that a merely visible influence in literature of scientific discoveries and theories is not a guarantee of more than a passing interest, a journalistic *mésalliance.* We maintain that the example we have examined of the contemporaneous interest in, and similarity of views on the nature of time, manifests the cultural impact of science on literature, which is a lasting influence—in this case a particular rather than a general influence. And it is the coherence of Eliot's philosophy and poetry that guarantees this cultural transformation.

PART IV
The Methodological Nature of Science

7

The Categorical Questions in Science

> Physical theory has two ardent desires, to gather up as far as possible all pertinent phenomena and their connections, and to help us not only to know *how* Nature is and *how* her transactions are carried through, but also to reach as far as possible the perhaps utopian and seemingly arrogant aim of knowing *why* Nature is thus and not otherwise. Here lies the highest satisfaction of a scientific person.
>
> —Albert Einstein

Introduction

That methodology that, in an earlier section, we argued to be the most pervasive of the cultural riches of science is, we repeat, different from the methodologies of the seventeenth century as well as those of the twentieth century. It referred neither to some putative way of assuredly arriving at new discoveries and the truth of nature; nor to correspondence rules for the ready appraisal and confirmation of theories. It had rather less strict, philosophical connotations. It referred instead to the scientific method of all the centuries, of quantification, experimentation, and generalization. In this sense methodology would be of some interest to the philosopher but even more to the historian who would find more substance in it. In the sense in which we shall investigate methodology in this chapter, more modern in its concerns, both history and philosophy are of importance. Indeed Kuhn, in his epochal book, has made it obvious that history is necessary for a meaningful appreciation of methodology. As for philosophy we can do no better than refer to Lakatos's paraphrase of Kant that "history of science without philosophy of science is blind,"[1] which was the way he introduced his own modern methodology.

Our earlier section on aesthetics implied a methodological

commitment to aesthetic criteria. This was true for certain scientists, notably Einstein, although more generally it was the case that methodologists often *justified* particular scientific *faits accomplis* by appealing to aesthetic criteria. Our tack in this section will be very different. We shall be concerned with a broader and more fundamental issue, which, at the same time, links up with a particular modern methodology.

The issue, which is not confined to modern science, may be most perspicuously introduced by referring to Galileo's asking of *how* questions, which has often been seen as the paradigm for scientific development, certainly when placed against Aristotle's failure in the asking of *why* questions. This is the starting point for this chapter in which we intend to present the uncontentious case that it is possible to detect three broad categories of questions that have been asked about science, these being *what, how,* and *why.* We say that the case is uncontentious because science has been discussed in terms of these categories before; indeed, *scientia* has been so discussed. These categories, however, might appear to be too broad and therefore to be of limited prescriptive value. However, very specific and narrow questions that might more readily be answered cannot be prescribed for the wide range and complexity of problems that do arise in the numerous areas across the whole range of physics. This is what we are interested in, and we shall have to be satisfied with the limitations of the broader questions. At the same time we shall incidentally show that there is some justification for them, some sense in which they are the inescapable questions.

Where our position differs markedly, and more contentiously from previous ones, is that we shall show how it is possible to understand those categories within the Kuhnian normal-revolutionary structure of science: the Kuhnian analysis is assumed to be valid rather than argued to be so. Placing our categories in such a framework will allow a more refined understanding of them, and we shall be better able to look, historically, at the solutions of previous problems from which we hope to (and shall) gain support for the refinement we have introduced.

We shall find that within normal science the important questions are *how,* although *what* is sometimes of (trivial) interest. Within revolutionary science *why* and *what* are of importance.

Furthermore, to the extent that the three interrogative categories are more fundamental than the Kuhnian structure, this structure may be regarded as gaining some support through any success that our case achieves. For it is not very difficult for us to accept the mere proposal that at the fundamental level of *scientia* (and therefore of science) such categorical questions are what confront us. This acceptance then implies the fundamental nature of methodology to the scientific enterprise, which is the important point for this chapter. If the most natural link of this general methodology is to the Kuhnian one, then automatically Kuhn's methodology acquires status.

The Historical Background

The ancient natural philosophers, particularly the Greeks, have been subjected, from time to time, to salvoes of insults. La Popelinière felt that "beyond beauty of language and quickness of mind, the Greeks do not seem to have merited as much praise as writers commonly give them."[2] Bacon, with his relentless empiricism, wrote in a more sharp-edged judgment that their "wisdom was fruitful of controversies but barren of works."[3] More particularly, the Greeks have been regarded as having entertained incorrect views about science, and they have consequently been blamed for the lack of progress in this field over a period of centuries.

Crombie, with an oblique thrust at the Greeks, accounted for the scientific revolution of the seventeenth century by arguing that it "came about by men asking questions within the range of an experimental answer by limiting their inquiries to physical rather than metaphysical problems."[4] Here, Crombie sharpens the Baconian view of the matter. The essence of Crombie's comment coincides with the view comonly held by many writers on *scientia* that, in the advancement of knowledge, asking the right questions is the most vital thing; it distinguishes genius from mere talent. Bohm, for example, asserts that "it is frequently realised that half the battle is over when we know what are the right questions to ask."[5] Bromberger goes further to say that "A clear mark of scientific genius is the ability to see certain well-known facts as departures from general rules . . . and the germane ability to ask why-questions that occur to no one else."[6] The defects of the Greeks therefore may be thought to be that they failed to ask the right questions.

The problem, however, does not necessarily disappear with this formulation for we are still left with the awesome task of prescribing what these questions should be. It might be that it is only from that embarrassing view of hindsight that we can detect what kinds of questions should have been asked; but, at least, they could be a guide for the future.

Now an apparent paradox of this formulation, if we follow Bromberger, is that Galileo is thought to have made significant advances in science specifically by asking *how* questions as opposed to Aristotle and the Greeks who asked *why* questions. This apparent paradox will only be resolved in the last section, but its emergence at this point allows us to introduce the *why-how* dispute in its historical context.

The *why-how* distinction has often been discussed in connection with the methodological innovations of Galileo vis-à-vis Aristotle. Dijkterhuis in his classic text, *The Mechanisation of the World Picture,* introduced Galileo's attack on the problem of falling bodies by noting that Galileo was fully aware of what he was doing in abandoning the old approach. He was not concerned with "the cause of motion even less so its purpose, . . . but exclusively [with] the way it takes place. His first concern is not with explanation, but with description."[7]

Furthermore, Crombie argued that Galileo wanted to show that the Aristotelian explanations "were . . . answers to the wrong questions," concerned as they were with the real essence of the physical world. "Galileo put aside all discussion of the 'essential natures' . . . and concentrated on describing what he observed."[8]

We can easily interpret this distinction between explanation and description as that between *why* and *how.*

More pointedly Burtt's enlightening work on modern science specifically referred to Galileo's abandonment of the final casuality of Aristotle in which "The analysis . . . [is] intended to answer the question *why* . . . rather than *how.*" He went on to say that with Galileo "it is the *how* of motion that becomes the object of analysis."[9] Indeed we already saw in Chapter 1 that one of Aristotle's most striking philosophical stances was in the asking of *why* questions.

The concern with arguing the case of Galileo's change in methodological orientation has continued apace in the book *New Perspectives on Galileo.* The introduction to that book reminds us that "for many years textbook accounts of Galileo's philosophy of science attributed to him the expurgation of final causes . . . : from now on science will deal only with efficient causes."[10] Pitt, in his article in the same book, specifically argues that Galileo worked within the Aristotelian framework that distinguished between *scientia quia* and *scientia propter quid,* that is between the *what* and the *why* of science. Pitt's view is that Galileo denied "the priority of *scientia propter quid* by relying on the descriptive prowess of a mathematical tool."[11]

It is also possible to document and therefore to emphasize the historical reality of the distinction by noting that Galileo, in his own time, had methodological opposition from Descartes who contrasted his own position by referring to Galileo's *Discourses concerning Two Sciences* as "merely seeking reasons for certain particular effects." In connection with Galileo's writings on the Lever "he explains very well what happens, but not why it happens, as I have done in my *Principles*."[12] There can be no doubt about the long-standing and continuing interest in the issue.

Pitt's comments do three things. They show that the origin of the distinction resides in Aristotle's philosophy of science. They show Galileo rejecting the *why* of science, but they also show him alternatively embracing the *what* of science through relying on the *how* that mathematics encourages. Thus is introduced the third category *what*. Thirdly, and related to the first point, this category is seen not to be new, and the manner of its introduction suggests that it might not be too distinct from the category *how;* this will be discussed in a later section. For the present we shall merely introduce the context within which *what* has recently been discussed.

This issue concerning the types of question that may be asked has engaged historians as well as scientists as we have implied before. This is not surprising if we see their task as not only describing but explaining historical events. Heller acknowledges the distinction made in science by remarking on Goethe's fear that science's successful "posing and answering questions about the 'How' of things, . . . might finally abolish in the world all creative interest in *what* this, that, or the other *are* and *mean*."[13]

In Heller's book also we learn of Toynbee's fear that his research project has been preempted by Spengler only to find that because there were no answers to the "question of *how* civilisations come into being that . . . there was still work for him to do."[14] Heller thinks that both types of questions are useful although the success of empirical science would seem to give priority to *how.* We note that he sees *the* important questions in science as *how?* For him "*what* a thing is raises a question categorically different from the question of how it came about;"[15] and in his own words "This state of intellectual affairs may have been brought about by the enormous birth-rate of modern science which has overpopulated our minds with most successful answers to questions of *How*, and left little breathing-space for questions of *What*."[16] We suggest that the last sentiment reflects the state of science but with the difference that most of the discussions have centered around *why* and *how* questions rather than around *how* questions alone, which is implied above.

The fertility of the distinctions and their relevance may be further adduced from the modern Hampelian[17] epistemological thesis that scientific knowledge can be analyzed into facts, laws, and theories corresponding to *what, how,* and *why*. The analogy is not perfect but certainly instructive as we shall try to show. Our immediate task, however, is to examine the source of the interrogative classification.

Aristotle and *Why* Questions

The origin of the issue we are discussing was noted to reside in the works of Aristotle, and it is for this reason that we shall look very quickly at his views on categorical questions. In his *Posterior Analytics* he argued that "Knowledge of the fact differs from knowledge of the reasoned fact."[18] He introduced the example of the "non-twinkling of planets" as fact, and that "they are near" as the reasoned fact.[19] He went on to say that knowledge of the reasoned fact or the *why* of knowledge is the more scientific "for grasp of a reasoned conclusion is the primary condition of knowledge."[20] More specifically we leave the burden with Aristotle because in his *Physics* he said that "Knowledge is the object of our enquiry, and men do not think they know a thing till they have grasped the 'why' of it."[21] He also gave concrete examples of his methodological position, which at the same time contains the seeds of the (modern) distinction between the technician and the scientist. To Aristotle "men of experience know that the thing is so, but do not know why, while the others know the 'why' and the cause. Hence we think also that the masterworkers in each craft . . . know in a truer sense . . . than the manual workers hence they know the causes of the things that are done."[22] Aristotle put himself firmly on the side of those who would seek scientific answers from nature that responded to *why* questions.

This distinction is connected to another that arose with Aristotle, for we can say that the world is made up of "substance" and "accidents," and nothing else besides. Whether we accept Aristotle's sophisticated view of substance, Newton's more commonsensical view, or Faraday's and Einstein's rather more tenuous views, it is of no moment. This is an allowable disagreement among philosophers. Whether we agree with the actual nine "accidents" of Aristotle might be of more importance but, here, only in matters of detail. We accept that Aristotle gave great importance to a discussion of the ontology of the world, and indeed both distinctions were referred to in Chapter 1 when we discussed Aristotle's views on *scientia*. This ontology is nothing more than our basic concerns with "what there is," which concerns are complemented and fulfilled by the epistemological task of grasping the "why of things." These concerns, however limited they might be in individual cases, are inescapable; they are the signposts of human existence. Aristotle classified these scientific questions as *scientia propter quid* and *scientia quia*.

The concern with *scientia quia* and *scientia propter quid* was kept alive for more than a millennium for we find Grosseteste in the thirteenth century phrasing his philosophy of science in the same way. Its continued vitality was assured because Galileo, who knew his Aristotle, chose to articulate his new method of attack within the same framework.

Categorical Questions and Kuhn's Methodology

We can now accept that over the centuries from Aristotle to the present time the questions that have been asked of science may be classified into the following three: *what, how,* and *why. What* and *why* generally lend themselves readily to much debate; and the substance of such questions, and the classification itself, have been frequently debated. If we regard philosophical questions as those that are endlessly debatable, and for the reason that philosophical discourse is largely an ingenious invention of new ways to say old things, then we might readily relegate *what* and *why* to philosophy and appropriate *how* to science. This would simplify our task; it would also gain support from the Galilean adoption of *how,* which (it might be argued) led to the escalating success of science ever after.

However, although Galileo followed a reasonable and tactically sound methodological path, it would be premature and simplistic for us to make the dismissive distinction intimated previously; for physics has a rich and unavoidable metaphysical component that thrives on theoretical entities, the source of much ontological scrimmaging. The method of physics is not limited to the ploy Galileo used in his kinematic researches. This brings us therefore to a refinement of the usage of these questions.

To begin with, it must be assumed that the questions to be asked of science are genuinely serious ones. "What is the composition of chlorine atoms?" "How does Mercury move around the Sun?" "Why does the atom not collapse into its nucleus?" These are not *normally* to be regarded as serious scientific questions at the present time: the answers are well known, and we are concerned with discovery rather than instruction. Stating our problem in this way allows us to introduce the major distinction between "normal" and "revolutionary" science according to Kuhn.

It will be recalled that Kuhn's theory of the structure of scientific revolutions separates normal, paradigmatic, puzzle-solving science from revolutionary, crisis-ridden science. In the former scientists do not need "novelties of fact or theory"[23] and so (according to the Hempelian correspondence) would not be concerned so much with *what* or *why*. This leaves them to be preoccupied with *how* questions. More positively, to the extent that normal science is puzzle solving, *how* assumes a position of major importance.

It is therefore to be argued that in the former, paradigmatic, regime we seriously (although not exclusively) ask *how* questions. In the latter, where we are at the frontiers of knowledge and metaphysics is unavoidable, we ask *what* and *why* as well. Thus, although we might relegate *what* and *why* to philosophy, we cannot divorce philosophy from science, as we have seen in earlier chapters.

The implication of the relationship between normal science and *how* questions is that if we are happily working within a paradigm we should address nature with *how* questions: methodologically *what* is often irrelevant and trivial, and *why* is trivial, as we shall show.

Normal science describes that activity that most scientists are engaged in most of the time. It is particularly noticeable that, whereas philosophers might disagree about Kuhn's analysis, most physicists who express an opinion agree with the Kuhnian distinction. It must therefore correspond to the perceptions of those whose opinions cannot be ignored. Normal science represents the consolidation that took place in the eighteenth century, of Newton's major discoveries; in the early nineteenth century, of

Young's and Fresnel's optical discoveries; in the late nineteenth century, of Maxwell's electromagnetic theory; and in the early twentieth century, from the 1920s, of Schrödinger's and Heisenberg's quantum theories.

Normal science represents the working out of all the implications of new concepts and theories with the help of new tools. In the way that the formulation of specified research projects for graduate students concentrates their minds and directs their efforts toward the acquisition and use of special tools, so on a larger scale for the wider scientific community, the arrival and acceptance of the quantum, wave-particle duality, and matrix mechanics engaged their interest and concentrated their efforts in the mastering of the new tools and the operation of the new paradigm.

The comparison is not precise for the (presently) fresh research students have no previous experience of operating actively within a paradigm. The comparison is valid in so far as it is freshly acquired tools that both (overlapping groups) must employ, in gaining mastery for the solution of problems.

Furthermore, in the narrower context, the majority of research students working in Atmospheric Sciences, say, are not fundamentally concerned either with the existence of ionized layers or with the correctness of Maxwell's electromagnetic theory; they take these things for granted. In the wider context, in the 1930s and 1940s, and later, scientists accepted that quanta existed, that duality was a reality, that Schrödinger's or Klein-Gordon's equations were good tools to use. They were not concerned with the questions *what* or *why*. To the extent that we accept that there is normal science that operates within a paradigm, and that the practitioners are involved with puzzle solving, then we can accept that they deal with *how* questions, which is a contention that even Aristotle might now support.

When we say that *what* is an irrelevant and trivial question within normal science, we mean both that the ontology of the science is well known and accepted, and that if such a question is asked it is merely as a preliminary to the major question *how*, or simply an alternative way of phrasing such a question.[24] When we say that *why* is trivial, we mean that it can invariably be translated into a *how* question. To understand fully what these qualifications mean, we should look individually at the three types of questions and how they are used; it is best to do this within the context of actual cases of scientific investigations. Within such cases, we shall argue for the relation between revolutionary science on the one hand, and *why* and *what* questions on the other.[25]

Case Studies: Normal Science and *How* Questions

Within the context of normal science, as defined, the major interest is in the techniques that support the particular paradigm. Technical expertise entails know-how so that technically competent (normal) scientists are those who get answers from nature to their *how* questions, consistent with their expertise.

In a straightforward and common case, electron microscopists might wonder what the contrast effects are on micrographs they are studying. It is a matter of *identifying* them as stacking fault, dislocation loop, void, or some other defect. In the most essential way the scientists are concerned with showing how they can use the armory of their paradigm to explain this one of a previously classified group of effects, within the particular context of their observations. The fundamental entities are not in doubt; the theoretical framework or explanation has been worked out. All that remains in the many instances, whose manifestations vary, is to employ the theory to explain their particularity. The questions that are dealt with are *how* questions.

Let us next take a particular historical case, that of pulsar research. There is a history of this work that stretches back to Zwicky[26] and his postulation of neutron stars, and to Oppenheimer[27] who helped to work out the theoretical behavior of these putative stars. To the extent that they were human, when Hewish et al.[28] looked at the jagged lines on their charts, they would have asked, reflexively, what they represented, what was the cause of their particular structure. They employed their acquired technique simply to establish the quiddity of these signs: they quite soon determined that they were neither interference nor scintillation effects.[29]

The history of the so-called discovery shows that there were two remaining candidates: white dwarfs and neutron stars, and the latter quickly won the day. The fact that the history and the paradigm existed, provided them with the initial task of *identifying* their object as one of the things already "known."[30] We see therefore that their *what* question was trivial, insofar as it was asking "which of the known objects is the source of this pattern?"

The major task from then on, has been to use the theoretical and experimental know-how to answer questions as to how particular pulsars arose, how their magnetic fields and their rotation account for observed phenomena. The question *why?* did not arise initially, and there was no need for it. Even long after, it arose trivially as an alternative *how* question.

I have argued that *why* is excluded and *what* minimized, reduced to the need to identify rather than to discover. Let us justify this statement in more detail. If we look more closely at the case of pulsar research, we find an initial concern with marks on a graph that were not unusual. They were obviously the result of rapid signals, and to this extent *what* was irrelevant; but their sources was uncertain. To the extent that it was necessary only to determine which of the well-known white dwarf or neutron star was responsible, they also knew *what* these signals might be. But to the extent that they needed to determine which particular one was the agent they might be thought to have been concerned with the question *what*. Such a *what* question is trivial despite the obvious fact that the observational data would be different for different kinds of stars. But it is trivial to the extent that what was then effected was an identification, rather than a discovery.

When we go beyond the initial problem of identification, we find various questions: why should the pulse-period vary? Or, more deeply, why is the neutron star's rotation slowing down? These are certainly not irrelevant questions. But they are not questions of genuine puzzlement; they phrase questions that are really asking how can we, with the knowledge at our disposal, explain these generally comprehensible phenomena. It is not as if scientists did not understand what the slowing down could mean. To this extent therefore they were concerned with attacking the problems in terms of *how;* and such were the answers they got.

Our final example of normal science is Lorentz's theoretical studies of electromagnetism. The central fact of the electromagnetic theories of Lorentz as well as of Maxwell and Poincaré was the ether. It was (presumed to be) the essential though embarrassing carrier of electromagnetic phenomena. One is tempted to compare the ether dogma with the circular dogma of ancient and medieval astronomy, although the temporal extent of their influence was quite different and the ether might be thought to have had a greater persuasive force because of its perceived *physical* necessity in contrast to the earlier dogma's *metaphysical* necessity.

The comparison can be maintained, however, because Lorentz's theory can be seen as a similar attempt to "save the phenomena," although with a certain inconsistency. It is to the extent that Lorentz's theory was saving the phenomena that it was concerned with *how* questions. Thus his full-blown theory set out to save the established concept of the (stationary) ether and Newtonian relativity; it indirectly accepted absolute space and time, while at the same time it abandoned Newton's third law.[31] Furthermore, Lorentz had in his grasp the new concepts of space and time, especially of time but cautiously accepted it only as local time, mathematically convenient. For he was not concerned with asking fundamental questions about these categories. His mind was not sufficiently open to seize upon these new concepts: he was working within a tradition, under the influence of a paradigm.

As Holton has remarked, Lorentz was attending to a very difficult problem with *conventional* tools. "At the time . . . the theory of the aether drag was in a quandary. To reconcile Maxwell's theory with the experiments on the aether was a great problem. . . . This was the problem to which H. A. Lorentz addressed himself."[32]

A clue that links Lorentz to a paradigm and reveals the nature of his questions was his use, admitted by him of various ad hoc hypothesis (reminiscent of Ptolemy's work) for the purpose of saving the phenomena. Lorentz was aware of the defect and of Poincaré's criticisms, for he relates that "Poincaré has objected to the existing theory . . . that, in order to explain Michelson's negative result, the introduction of a new hypothesis has been required."[33] We shall elaborate on his procedure.

Lorentz initially set out to show how Fresnel's (first order) dragging coefficient could be accommodated in his theory of electrons, and he succeeded. But Michelson's experimental null results for second order effects were a problem. Lorentz supports our major contention about his work by revealing that "This experiment has been puzzling me for a long time, and in the end I have been able to think of only one means of reconciling its results with Fresnel's theory."[34] With the supposition that molecular forces behave in the same way as electromagnetic forces, he then introduced his "Lorentz contraction."

His problems did not end here, however, for he had to account for a host of other optical and electromagnetic phenomena in the moving frame of the earth through the ether. He did attempt a comparison of laws in the moving and in the rest frames but his method then was to see what transformations were required to eliminate the velocity of the moving frame.

It is clear from the account of Lorentz's approach that he was hewing to the traditional line, attempting to solve puzzles as they arose, employing the conventional concepts and tools. His purpose, for example, of "reconciling" Michelson's results with Fresnel's was basically and grammatically one of asking *how* these results could be understood within his theory rather than *why* the solution was as it seemed. It was a technician's job. He did not go deeply enough to ask the searching fundamental *why* questions: he was an extraordinarily gifted technician, trying to show how things might be done. It is true that H. A. Lorentz was unlike many other heroes in that he was less philosophically minded and temperamentally more cautious. Einstein was the complete opposite, and it is to the point to recall that even after Lorentz had to admit that Einstein's theory was "correct" he wondered whether Einstein had not gone too far. Lorentz preferred to "decorate" an existing system incorporating the ether than to "put forward a personal view as self-evident," as Einstein had done. Einstein had (revolutionary) daring, and Lorentz indeed admitted that they did physics rather differently, for where he was prepared to deduce principles, an eminently constructive and paradigmatic approach, Einstein simply assumed these principles.[35]

Case Study: Revolutionary Science and *Why* Questions

Lorentz had achieved enormous success, but fundamental crises still remained within the body physical. The situation was ripe for one of Einstein's particular mind and temperament.

Now major revolutions in science, as the name implies, go to the very roots of the prevailing theories, turning them over because of dissatisfaction, because it is not accepted that things should be as they are. The motivating questions in such a situation are *why* questions although in the establishment of the new vision, *how* questions also become relevant. It is our purpose in this section to show first that Einstein's work was revolutionary in Kuhn's sense. Einstein arrived at his theory with its astounding power, by boldly accepting the reality of Lorentz's local time, and thereby, the paradigm was changed: this change and in fact much of the theory arose from (fundamental) *why* questions.

To begin with, we have Einstein's opinion on the *how-why* methodologies. He believed that "Physical theory has two ardent desires, to gather up as far as possible all pertinent phenomena and their connections, and to help us not only to know *how* Nature is and *how* her transactions are carried through, but also to reach as far as possible the perhaps utopian and seemingly arrogant aim of knowing *why* Nature is *thus and not otherwise*. Here lies the highest satisfaction of a scientific person." This passage reflects Einstein's passionate relationship with science. But it is also clearly stating that the questions that physics asks are *how* and *why;* and that it is answers to *why* questions that take us to the limits of knowledge and thus provide the greatest satisfaction. The quotation, we suggest, reflects Einstein's own position.[36]

Einstein refers to "the . . . seemingly arrogant aim of knowing *why* Nature is *thus and not otherwise*" and he would have been thought by some to be arrogant. Thus one of his earliest teachers remarked that what was wrong with him was that "One can't tell you anything." As Holton described him, he seemed to possess an "unreasonable degree of optimism about his own mission . . . that appear[ed] to others at times to be egocentric obstinacy."[37] The serene confidence in his ability is revealed very simply in a letter written in his old age, to Born, about Whittaker's denying him sufficient credit for the theory of relativity: "At any rate I found satisfaction in my efforts."[38] The passion and the satisfaction are glimpsed elsewhere in the literature, for example in one of his interviews with Shankland when he refers to "the seven years that relativity had been my life."[39]

We can therefore take the additional step and accept that in Einstein's characterization of Physical Science, he characterized his own work and that, supremely, what he tried to do was to discover "*why* Nature is *thus and not otherwise*." To begin to take a closer look at his work, in this way, we recall his youthful thought experiment with a light beam, which led him to accept that the velocity of light must be the same for all observers. The questions that this acceptance provoked were not how he could account for it within the existing framework, but *why* it should be so. This is supported by the oft-quoted passage concerning his views on the origin of theories where he admits that "By and by I despaired of the possibility of discovering the true laws by means of *constructive* efforts . . . I came to the conviction that only the discovery of a universal formal principle could lead to assured results."[40] He was not going to be engaged in a technician's job but in a philosopher's in which the ultimate secrets of nature would be sought. Finally he refers to the example of Thermodynamics, which he followed, where "the laws of nature are such that it is impossible to construct a *perpetuum mobile*."[41] With this inspiration he proposed the formal principle of relativity that provided him with the answer to his youthful question *why* nature is thus and not otherwise.[42]

When we look, specifically at his original paper we find him, in the very first paragraph, raising fundamental questions about symmetry: Why should an asymmetry show up in the theory of electromagnetic effects between conductor and relatively moving magnet? No constructive theory is going to solve this. The solution is to be found in that universal principle that he was setting out to establish.

We could similarly analyze his introduction of the convention for simultaneity, which led eventually to revolutionary views about time, and find him again asking *why* question. It will be significant to relate for the support of our general case that Einstein once said "that he was brought to the formulation of relativity theory in good part because he kept asking himself questions concerning space and time that only children wonder about."[43] Anyone who has had contact with children need not be reminded that they relentlessly, and almost exclusively, ask *why* questions.

It is now perfectly possible to maintain that Einstein not only philosophically subscribed to the *how-why* distinction, but actually used the categories. More important, the evidence strongly suggests that he associated important revolutionary work with *why* questions. We can let this case of Einstein's theory of relativity stand for the general point that we wish to make, especially when it is seen against the background of Lorentz's work. We shall only add that it is not too difficult to see some of Balmer's and Ritz's work as similar to Lorentz's in that they were attempting a patchwork job, trying to show *how* spectra could be understood. De Broglie and Schrödinger, on the other hand questioned why nature should be as it seemed, asymmetrical in its manifestations of particles and waves. The conviction was, as it was for Einstein, that nature was not asymmetrical. Hence, we arrived at Schrödinger's theory.

It has not been our intention, as indicated before, to exhaust cases, but to make is plausible by argument and by examples that our thesis is supportable. The main component of our thesis and our argument so far has been that *how* questions activate normal science, and *why* questions do the same for revolutionary science. It is only left to consider to what extent the categories *how* and *what* enter into revolutionary science.

Case Studies: Revolutionary Science and *What* Questions

What the story of the origins of relativity theory shows is that Einstein persistently questioned the status of time and space as they were then accepted. What he showed in his defintion of simultaneity, in his original paper, was *how* we could understand his new vision of time. Whereas it was the sharp *why* questions that led to the convictions, indeed to the insights into nature's reality, it was the asking and answering of *how* questions through which those insights were realized. If we extended the examples to general relativity we see that among other things Einstein was asking why should there be any special (inertial) frames of reference. His conviction was that there should be none. It was through the extension of the Minkowski formulation that he showed us *how* such an insight could be understood.

In any case once a breakthrough has been achieved and the new view accepted, we are now working within a (new) paradigm wherein *how* questions are the significant ones.

To a lesser extent *what* questions are (trivially) relevant as we saw before. But there is a final distinction to be made between such questions and those similar questions that help to introduce and establish the new paradigm. We shall therefore, finally, argue for the existence of *what* questions within revolutionary phases of science, which takes us into the area of the ontology of science.

What is there? Democritus and others gave the answer "Atoms and the void." Whether it is Democritus or the "man in the street," the question is the most fundamental one, concerning the furniture of the universe and, more intimately, the objects of our experience. Even if the man in the street does not question the nature of these objects he cannot avoid identifying them, at least.

For the scientist there are empirical and theoretical entities— atoms and force for example, which, however much their existence is debated, are accepted by philosophers within the operations of normal science. But precisely because the existence of these entities is endlessly debatable the questions concerning them, *what* questions, are quite often at the forefront of revolutionary science.

This characterization of *what* questions might seem contradictory or paradoxical, but it merely indicates the relative instability of ontological elements as we shall argue.

Now, one of the reasons why many of these entities spark regular debates is that they are not, in the most obvious sense, perceivable; the elementary particles exemplify this. Another reason is that they form part of perennial dualities so that in any age there will be some who adhere to the one and others to the other. These dualities represent the inescapable ways in which we can perceive the world. Examples of them are continuity-discreteness, permanance-flux, substance-form (accident). Particle-wave and mechanism-field are just derivatives of these dualities. There is no way, it seems, for the various ontological issues to be permanently settled. There have been and will be alternating dominance of the one, and the other, and we have seen that this is so in the particular case treated in Chapter 2.

We shall therefore look briefly at historical examples in which *what* questions played a significant part. The seventeenth-century revolution is usually seen as culminating in Newton and especially in his *Principia*. However, it is not denied that a major aspect of this revolution was his experimental researches in optics, which helped to establish the power of the experimental method; Newton himself was forever glancing back to the work he had done in optics. In the most obvious way he set out to discover what light was or more particularly what color was. He discovered that it was not something separate from light; and certainly it was not something that was imposed by a medium. To the extent that the nature of light has always been in the forefront of physical theory and in the philosophy of science, this clarification was of major significance.

However, the example, important though it is, does not fall within the category of those endlessly debatable issues. An example of such is whether the universe is a plenum or a collection of discrete atoms. Democritus proposed the atom and Aristotle denied it; Galileo supported it and Descartes demurred. The fundamental basis of Newton's revolutionary work was that the answer to the question of what there is may be given as (atomic) particles and a force (between them). Thus was introduced force as one of the existing things of the world consistent with the acceptance of the atomic model. The power and success of Newton's work imposed this world view upon the community so that Dalton's important work could be seen as its final verification, certainly of the atomic model, and thereby of force.

This was not the last word, however, for Faraday who spent his life searching for a world view specifically sought an answer to the question bandied about by Newtonians and by Cartesians, whether the world consisted of atoms and force or of some plenum of vortex-fluid. What he did was to suppose that the forces themselves are the sole physical substance.

It must be emphasized that this view did not arise from Faraday asking *why* or *how* questions. He had been influenced by the philosophy of Kant and Leibnitz through Oersted to seek for the unifying substratum of the world, whatever it might be. It was after he established to his satisfaction the existence of the field that he tried to use it to explain how various phenomena could be understood.

The explanatory power that his theory had, the influence on Maxwell's electromagnetic theory, and the support of Hertz's experiments speak for the way in which the asking and answering of *what* questions can change the physical landscape, and revolutionize the world view. The twentieth century saw the amplification of this view through the field theory of Einstein, which supplanted Newton's atomistic gravitational theory, and we are still in the grip of such theories. *What* questions, it may be argued, play no less a role than *why* questions in the revolutionary stage of the development of scientific theories.

Conclusion

If it is true that *how* questions are of major importance in normal science, and *why* and *what* questions in revolutionary, how is it that much of the critical discussion of the development of science has centered on the methodological path pursued by Galileo as opposed to Aristotle, the former highlighting *how* questions as opposed to the latter's *why* questions?

We must answer this question if we wish to maintain our thesis, for there is no doubt that Galileo chose that particular path: the internal evidence and the critical accounts show this. There is no doubt either that the path he chose led to important developments. There is a conclusive answer, we suggest, that consists of a complex of reasons. The fact is that Galileo has been seen by critics to be central to the revolutionary science of his century, and one of the important things he did was to establish the kinematic equation for free fall on the declared basis of his "new" method. It has therefore been irresistible to associate the method with the revolution and because that revolution achieved so much, to associate the method with *all* revolutions. In fact, Galileo's kinematic equation can neither gain the significance and importance of being derived from experiment, nor can its importance be sustained by the claim of originality. What it did, which is not to say it was not important, was that it interpreted the work of Oresme[44] by showing that free fall was such a motion as Oresme had described.

It must not be overlooked, and it is certainly not being denied, that Galileo cleared an impasse by his methodological choice. But the point to grasp is that it was a good tactical move for *Galileo*. As Dijkterhuis commented: it is a "restriction he imposes on himself, not upon science";[45] this is supported by Galileo himself. He had chosen to ask *how* rather than *why* in relation to the phenomenon of motion; that is, he had tried to describe the *how* of motion rather than to enquire into the *why* or causes of motion—that is, he chose to establish a kinematics rather than a dynamics. Dijkterhuis went on to say that "As soon as it is accurately known how bodies fall . . . then perhaps more probing minds will be able to arrive at a profounder insight into the nature of the fall and its laws."[46] But thus, from this position, Galileo's work supports both our thesis and our view of his contribution. For, in the first place, it is accepted that it is dynamics that will yield the ultimate breakthrough and that the dynamical questions are *why* questions. Secondly, it is clear that the breakthrough was still to be realized. All of this is consonant with the historical circumstance of Newton's dynamics, and the interpretation that it was through his work that the revolution was consummated.

In the light of all this we can now resolve the apparent paradox referred to in the introduction. It is not a paradox: it is a misunderstanding both to say that Galileo showed the way to the fundamental advancement or revolution in science by means of *how* questions; and, by this yardstick, to castigate Aristotle for impeding science by asking the wrong, that is, the *why* questions. Both the source of the seventeenth century revolution and the errors of Aristotle lay elsewhere. For Aristotle wrote at a time when science was *scientia* and the predominant and important questions that were asked were philosophical, for which *why* questions are of eternal relevance. Criticizing Aristotle for asking the wrong questions is therefore unfair. As for Galileo, we have made the point that his choice of *how* questions was a methodological ploy, necessary at the time to achieve scientific progress. *Scientia* had now become science. Yet he was aware that further progress needed to be made, and if this were to be achieved, *why* questions would have to be asked. Undeniably science's progress can be inspired by the asking of both *why* and *how* questions, as well as of *what* questions.

8

Pulsar Research as Normal Science

> Radio telescopes are only the most recent examples of the
> lengths to which research workers will go if a paradigm as-
> sures them that the facts they seek are important.
>
> —T. S. Kuhn, *The Structure of Scientific Revolutions*

Introduction

This chapter sets out both to recall the scientific activity that
preceded and accompanied an episode in the recent history of
science, the discovery of pulsars, and to reveal thereby that this
piece of contemporary research is simply part of a continuous
thread that can be woven through the literature. We shall also
argue that this research satisfies the theory of scientific development, via normal periods, as propounded by Kuhn. The detailed
argument for this view will be taken up later in this chapter; it is
enough to argue at this stage that continuity is aspect of normal
science.

Normal science is opposed by Kuhn to revolutionary science. It
is the science that is usually practiced by identifiable groups of
scientists, operating within a paradigm, or, as amended by Kuhn,
in a disciplinary matrix, and in a tradition of shared examples.
The major aspect of this normal science is that it entails normal
measurements, and problem solving; and it is cumulative. Occasionally within each paradigm crises arise, a revolution takes
place, and a new paradigm is established. However, the revolutions are not to be regarded exclusively as the type like Relativity
and Quantum theories; they are generally more frequent and
smaller.

The choice of pulsar research for investigation arose partly
from the circumstance of its being a very active area of research
receiving publicity, which gave it the aura of a major discovery. We
shall see that, on the contrary, what was achieved was really the

173

detection of neutron stars. This fact supports the ascription of normal science to pulsar research.

Furthermore, if pulsar research exemplifies normal science, it should also support our methodological thesis about *how* questions as we presented the case in the previous chapter. We should therefore expect the account of pulsar research to reinforce that methodology by showing that progress was indeed achieved by the asking of *how* questions. The case for pulsar research being normal science was *presented* in the previous chapter; in this chapter the case will be *argued* in some detail so that, again in a major section, we have a particular case of the more general one presented in the previous chapter.

Background: Prewar Years

The central assumption of the widely accepted model of pulsars is that they are rotating neutron stars. This takes us back to the early 1930s when Chadwick discovered the neutron, and soon after when Zwicky and Baade boldly predicted the existence of neutron stars. From their studies of the energies involved in such phenomena as novae and supernovae (which they first named) they were led "with all reserve . . . to advance the view that a supernova represents the transition of an ordinary star into a *neutron star* . . . such a star may possess a very small radius and an extremely high density." In the context it was a bold suggestion although cautiously expressed. They went further in suggesting that supernovae might be the origin of cosmic rays. However, there were no theoretical formulations of these ideas, and so they remained vulnerable to indifference or facile attack. For example, Cernuschi[2] could seriously challenge the notion that neutron star formation *could* lead to the observed explosions.

What was required was a comprehensive theory for such a state of matter that was highly condensed and at the same time massive; such a theory began to develop in the 1930s. Eddington had suggested that at the high densities envisaged ionization would play a predominant role. It was then found that the known, highly condensed white dwarfs must have their electrons, which were detached from their nuclei existing as a degenerate gas in the sense of the Fermi-Dirac statistics.[3] At even higher densities it might become impossible for protons and electrons to exist separately and the resulting interaction could then lead to a degenerate neutron gas, in which there would be a pressure associated with degenerate states. Chadrasekhar,[4] with an appropriate equation of state, found an upper limit for stellar masses that could gravitationally support electron degeneracy pressure; this was 5.75 $M_\odot/\mu_e^2$ where $M_\odot$ is the mass of the sun and μ_e is the number of electrons per proton mass. For these white dwarfs the mass is 1.44 $M_\odot$ on the assumption that $\mu_e = 2$, which means both that the core hydrogen has been exhausted and that complete degeneracy exists. Development along these lines took place in natural harness with a theory for stellar evolution, the theory being supported by an explanation for the origin of stellar energy.

Much of the work at the time was exploratory and tentative, as might be expected. Zwicky himself had gone beyond the qualitative assertions about neutron stars and had arrived at values for their density and radius, namely $\rho = 10^{14}$ gm cm^{-3} and R$= 10^6$ cm but his subsequent work *On the Theory and Observation of Highly Collapsed Stars*[5] did not really give us a theory but mainly extended arguments for the transformation of an ordinary star to a neutron star via a supernova explosion. For example, he argued that the neutron star hypothesis could explain the observed redshifts of permanent features of supernovae if they were gravitational in origin. Zwicky was aware that an equation of state for a neutron gas was required, but he failed to produce one. What he did produce was a rather wide range of values for the mass of a neutron star, namely 32 $M_\odot > M > 0.1 M_\odot$.

Landau[6] took up the concept of a stable aggregate of neutrons in suggesting that all stars would necessarily have such a core. He also went on to consider mass-stability conditions for highly condensed objects, realizing that electron and neutron degeneracy pressures in opposition to gravitational pressures could lead to stable states different from an ordinary star. The lower and upper limits that he obtained for a neutron core were 0.001 $M_\odot$ and 6 $M_\odot$. Gamow and Teller[7] rejected the neutron core hypothesis on the basis that it would lead to much too high a stellar temperature. (We note that Gamow had earlier suggested that *in sufficiently massive* stars, after all thermonuclear sources of energy were exhausted, a neutron star might be formed.) However, the point to make is that the calculations of mass limits were dismissed by Oppenheimer et al. who found the range to be too wide even if more than the core were regarded as an aggregate of neutrons.

Oppenheimer and Serber[8] arrived at a lower limit of 0.1$M_\odot$, but argued that the limit would vary according to the type of nuclear force assumed. At worst it would be $M > 0.03\ M_\odot$. Oppenheimer and Volkoff[9] considered the possibility of a neutron core but dismissed Landau's limits, the lower because he had chosen the wrong composition, oxygen, and the upper limit because he had employed Newtonian gravitation. For such highly collapsed states, general relativity was felt to be indispensable. Using this theory they found static solutions for neutron cores employing an equation of state for a cold Fermi gas, which Zwicky had failed to do. This equation was given, for pressure and density, in a parametric form, and thermal energy and interparticle forces were neglected. Whereas Chandrasekhar had earlier found an upper limit for electron degeneracy pressure, Oppenheimer and Volkoff found a similar limit ($M < 0.7\ M_\odot$) for the higher densities and the neutron degeneracy that would exist in neutron stars. The range of masses that Oppenheimer et al. found was $0.03\ M_\odot < M < 0.7\ M_\odot$.

For $M > 0.7\ M_\odot$ there were no static solutions. What Oppenheimer and Snyder[10] did, therefore, was to seek the nonstatic solutions of the field equation of general relativity for massive stars. Whereas Oppenheimer and Volkoff had found that indefinite contraction might occur for very massive bodies, Oppenheimer and Snyder worked out the details of such an occurrence. They discussed the probable effect of deviations from a Fermi gas model ad concluded that for a star greater than a few solar masses indefinite contraction would take place after exhaustion of the thermonuclear source. In their solution they ignored radiation pressure, but its inclusion could be shown to have little effect on the predicted catastrophic phenomenon, which was subsequently labeled the "black hole."

Oppenheimer et al. had clearly furnished us with an appreciation of the immediate provenance of neutron stars. We also gained from the related activity of the period some understanding about nuclear energy sources in stars. On the first point, if we look forward to the discovery of pulsars as neutron stars, we can see how large a part nuclear and statistical theories play in the physics of highly condensed matter. Oppenheimer and Volkoff's work shows the importance also of general relativity, and Zwicky himself in his 1939 paper had stated that "The hypothesis of the formation of neutron stars would run into serious difficulties if one should attempt to retain the classical theory of gravitation." His explanation of supernovae redshifts as gravitational subscribes to the same theory. The comments show us how well laid were the theoretical foundations of pulsar research.

As far as the details of stellar evolution along and off the main sequence are concerned, the mechanism behind Gamow's and Bethe's theories take us a long way. Gamow[11] was the first to elaborate on the assertion that thermonuclear reactions were the source of ordinary stellar energy. He regarded the H-He reaction as the major energy producing source in stars. Bethe,[12] however, showed quite authoritatively that the CNO cycle was important, especially for ordinary but more massive stars. Hoyle's postwar theory of nucleosynthesis rests upon the same premises. We shall see, indeed, how much of the postwar astrophysical work was foreshadowed by work during the 1930s.

Background: Postwar Years

Oppenheimer and Snyder's paper was published on 1 September 1939. The war arrived and Oppenheimer became involved in other matters. Interest in neutron stars as such was not great during the immediate postwar years, with the work of of Zwicky, Oppenheimer, and others being almost forgotten. It was radio astronomy, for one, that was creating great excitement, but the techniques and theories that developed around this subject eventually opened up avenues to the detection and understanding of such stars. Hoyle played a significant part in this work.

Taking up one strand of the prewar story, Burbidge, Hoyle, et al. (1957)[13] provided a satisfactory theory for the first time of the synthesis of the elements inside stars that had traveled most of their evolutionary path. In their paper they gave the various astrophysical origins for different groups of elements. In 1960 Hoyle and Fowler[14] concentrated on those elements that, according to the earlier paper, would originate in supernovae, stars that manifest an enormous release of energy. This orientation toward high-energy astrophysics was part of the existing attempts to explain the origin of such massive systems as radio galaxies; it also provided the right direction for the investigation of those other energetic radio sources, quasars, which were first accepted as such in 1963. Hoyle and Fowler recognized two types of supernovae, I and II, partly related to initial mass of the order of $M_\odot$ and 30 $M_\odot$, respectively. Supernovae of the first type acquire relatively high degeneracy pressures, which are responsible for triggering explosions. These same pressures lead to a stable configuration for these stars: electron degeneracy pressure leads to a white dwarf for $M \simeq M_\odot$, and neutron degeneracy pressure to a neutron star for $M \simeq 2\text{--}3\ M_\odot$. These limits $M_\odot < M < 3\ M_\odot$, imply that evolution can take place beyond the white dwarf stage but will be halted before the the black hole stage is reached. Incidentally Hoyle and Fowler found that μ_e could be as much as 2.23 so that the critical white dwarf mass would be 1.16 $M_\odot$.

For supernovae of type II implosion leads to high enough temperatures for initiation of the explosive reaction: $Fe^{56} \rightarrow He^4 + 4n$. Implosion proceeds under overwhelming gravitational forces, and for a mass $M > 10\ M_\odot$ there is catastrophic collapse into a black hole. These massive stars do not normally develop densities through which high degeneracy pressures could halt such a collapse. Apart from reestablishing the theoretical framework for the understanding of neutron stars Hoyle and Fowler gave us more than a merely plausible connection between neutron stars and supernovae.

Hoyle et al.[15] returned to the subject in 1964 through their paper *On Relativistic Astronphysics*. They were able to show that, for highly condensed objects, the inclusion of radiation pressure, which Oppenheimer and Snyder had ignored, would lead to a speeding up of the onset of catastrophic collapse. Hoyle et al. advanced beyond the achievements of the 1930s by considering the effects, also, of nonzero temperatures, and of rotation. They were much concerned, as we shall see, with the possibility of a collapsed object acting as a continuous energy source. They considered, in particular, the problems of the origin of, and the activity within, the Crab nebula. The argument was that a supernova explosion would bequeath a remnant, and it was obvious from the moving wisps within the nebula that this remnant was very active. The point was that there had to be a coexisting source for the observed activity. Hoyle et al. thought it unlikely that the source would be a white dwarf because it would not satisfy the observed luminosity. Neutron stars seemed more promising, but they argued that it was improbable that a single neutron star was responsible for the nebular activity.

A particular problem that Hoyle et al. considered in 1964, as we saw, was that of a rotating object in a supernova explosion. Such a problem had not been solved within the necessary general relativistic formulations. However, Hoyle et al. were able to suggest that for $M > M_c$ (where M_c is the critical mass for formation of a neutron star) a rotating object would break up, perhaps into two or more neutron stars that would then act as sources. What was significant and was clear to them was that a highly condensed mass could act (also) nonexplosively as an energy source, and that such was the source of the activity in the Crab nebula. One important implication was that the neutron star would also be rotating.

There were other scientists whose astrophysical work readily incorporated the concept of a neutron star as an explanatory model. The discovery of galactic sources of x-rays in Scorpius and in the Crab nebula—for example, in 1962/63—quickly found these sources being explained in terms of neutron stars. Although this explanation could not be maintained in the face of more careful study—the angular size of the sources, from 20″ to 2′ of arc indicated that they were too large to be neutron stars— yet the mere resort to such stars was a clear indication of their credibility. Hoyle, Narlikar, and Wheeler (1964)[16] commented that "The recent detection of x-rays from the region of the Crab Nebula and from a discrete source in Scorpius . . . has much increased the interest in neutron stars." They themselves went on to propose that the radial pulsations of such stars might play an important part in supernovae and in their remnants. Meltzer and Thorne[17] considered the damping of neutron star pulsa- tions, which they, also, felt might be responsible for supernova activity. However, for the particular damping process considered they found that only if the efficiency of its energy conversion were greater than 30% could the pulsations be the source of the light energy observed in the corresponding supernova.

Finally, both Wheeler[18] and Pacini[19] in 1966 and 1967 consid- ered more explicitly than Hoyle et al. that *rotating* neutron stars might provide the energy for supernova remnants. Pacini felt certain that a dense stellar core is left behind after an explosion but that the ensuing pulsations of the core would be quickly damped out, as we have noted. He therefore resorted to a magne- tized oblique rotator model for the postulated neutron star in the Crab nebula, and showed that large amounts of energy and mo- mentum could be released from such a star into the rest of the nebula, in the form of very low frequency electromagnetic waves.

We see very clearly at this point that all the speculations and half-formed theories of the "thirties were acquiring a relentless cogency from the theoretical activity of the sixties." By 1967 the stage was neatly and fully set for the discovery of pulsars and their elucidation as rotating neutron stars.

Pulsar Discovery and Pulsar Research

The blueprint for neutron star detection was not followed. Despite their credible association with supernovae, and their use as explanatory models from time to time, it was not felt that these stars, would, generally, be easily detected. It is not surprising therefore that they were discovered quite accidentally: Hewish et al. were using a modified experimental arrangement in a study of quasars by scintillations. What they discovered was the very precise repetition rate of radio pulses. As such these pulses were surprising but not anomalous, and they clearly required interpretation. With varying degrees of seriousness, several possibilities as to their origin were considered, terrestrial and extra-solar. Hewish et al. finally narrowed them down to "the stable oscillations of white dwarf or neutron stars."[20] The elimination proceeded along rational lines, but it is interesting that neutron stars turned up again, maintaining their growing role as explanatory models for celestial phenomena. Validation was widely sought and fairly quickly achieved.

The pattern of activity that ensued the discovery of pulsars was fairly standard. We saw attempts made to carry out detailed analysis of the data in hand, to obtain other kinds of information on the radio sources and to establish a theory that would account for all the observational features although not necessarily in this tidy sequence. In these attempts, the distance, size, and period of the sources were more accurately measured and pulse structure and amplitude variations analyzed. At the same time determinations were made of the kind and degree of polarization in the radio pulses, and surveys were carried out to obtain identification with optical sources, and to ascertain the galactic distribution of the sources and pulsar-supernova coincidences. Other features were also observed, for example, the generally continuous lengthening of inter-pulse periods and, in particular, the rather high rate at which the Crab pulsar was slowing down.

Of course, what was most keenly sought was a model that could consistently explain all of the phenomena. It was necessary to explain the clock mechanism, the radiation mechanism, and the beaming mechanism. During 1968/69 various journals, with *Nature* in the lead, published a host of speculations and theories employing all the models we have hinted at; these included Gold's model of a rotating neutron star, which was published quite early, six months after the first communication about pulsars. It is of interest to note that, nevertheless, white dwarf and other models were being published even after Gold's model seemed the most likely candidate.

A reconstruction of the discussion that took place before Gold's model became generally acceptable will be useful. The discussion entailed, for example, deductions about the size of the source ($\sim$ 10 km). This value led to a brief for white dwarf or neutron star and a galactic residence. Such stars might have circumstellar plasma but the precision of the timekeeping ruled out the plasma as its source: the accuracy of the measured period was $\sim$ 1: 10^8. Soon after the initial publication on pulsars rotation was introduced as a possible source for the timekeeping: rotation of a star or of a planetary system of neutron star and satellite. The latter was dismissed by Pacini and Salpeter,[21] who showed that the mass of such a satellite would be too small. White dwarfs were also dismissed. They were too massive to rotate fast enough, of the order of milliseconds, without rupture, and their period of pulsation would be appreciably larger than the observed values. Meltzer and Thorne's theory gave 8 s for the pulsation period, and even when an error in their assumptions was pointed out the period did not fall much below 2 s; the range of observed periods was, at the time, 33 ms to under 2 s. This circumstance left us with the neutron star. Gold[22] referred to the virial theorem that the lowest mode of the pulsating period should be $\simeq$ the maximum rotational period that the star could possess without rupture. This seemed to exclude pulsations. We were thus left with Gold's model of a rotating neutron star that could satisfactorily explain the clock mechanism as well as some details.

Gold's model of a pulsar, in detail, was that of an axisymmetric, rotating neutron star in which a huge magnetic field ($B \simeq 10^{12}$ gauss) had been frozen, and to which was attached a corotating magnetosphere. The size of the star would limit the pulse width, and its rotation period would be the same as the pulsar period. Thus both pulse width and period were explained by this model. For the radiation mechanism it was proposed that a surface instability on the star surface led to relativistic acceleration of particles across the corotating magnetosphere. This rotation could take place only out to a radius given by $r = c/\Omega$, where Ω is the angular velocity of the star and c the velocity of light. The impossibility of crossing the velocity of light barrier would in some way lead to radio radiation at that barrier. It would take place tangentially within a small angle, and the accompanying radiation reaction together with the loss of relativistic gas escaping from the magnetosphere should lead to a braking of the rotating system and a gradual lengthening of the pulse period.

Acceptance of Gold's model grew mainly out of the success of this prediction of the change in period; and the model could also be exploited for a neutron star–supernova connection. Gold had indeed quite early suggested that shorter-period pulsars should exist and be associated with young supernovae. This prediction was also successful as the connection was supported both by the discovery of pulsars in the supernova remnants of Vela and of the Crab nebulae,[23] and by the (young) Crab pulsar having (at that time) the shortest period. There were other aspects of Gold's model, however, about which reservations were expressed.

For example, Gold presented us with more of a model or a scheme than a theory, and despite some amount of quantification to show that cosmic rays may have risen from rotating neutron stars' losses in their early, faster spinning regime, he did not really establish, beyond a kind of rhetorical plausibility, how we could understand the radiation mechanism. Furthermore, the observation of optical pulses from the Crab nebula with the same repetition rate that the radio pulses possessed suggested that some theoretical modification was required. Many theoretical astrophysicists obliged, but despite their continuing efforts we still did not have any satisfactory explanation of the radiation mechanism.

The fact that the theory of pulsars was incomplete (in 1972) is unimportant for our purpose. What is important is that Zwicky's hypothesis of a neutron star and its supernova connection had worked its way via both the theories of Oppenheimer et al., Hoyle et al., and Pacini, for example, and the observations of Hewish et al. and others, to the current certainty. Although the research up to the 1960s did not set out to discover neutron stars and their associations according to the work of those mentioned, this is exactly what transpired. There is undoubtedly a retrospective continuity in the work, and pulsar research can thus be seen to have an ancestry that goes back almost five decades.

That pulsar research encompasses these predictions and theories lends support to the contention that what was achieved was more the detection of neutron stars than their discovery; and this point of view reinforces the validity of labeling pulsar research as normal science.

More detailed consideration of the theoretical work done in pulsar research will show that it has been mainly a puzzle-solving activity existing within a single paradigm. This partly explains why the rotating neutron star mode created no *shock* in the relevant research communities; and, after all, we saw that Wheeler and Pacini had introduced a similar model less than two years before pulsars were actually discovered. The ease of assimilation is a function of all the (research) characteristics discussed and is a persuasive argument that pulsar research has indeed been normal science in the context of Kuhn's theory. Our next task is to flesh out the account of previous sections by giving a detailed account of the methodology of science at work.

Pulsar Research as Paradigmatic

It is important for an understanding of the nature and effects of pulsar research to locate its milieu. In any case, the milieu is the base of the paradigmatic nature that we wish to establish. The question as to whether pulsar research is normal or revolutionary science can then be addressed to the appropriate practitioners. Kuhn's views on the matter are quite clear:" A paradigm governs, in the first instance, not a subject matter but rather a group of practitioners. Any study of paradigm-directed or of paradigm-shattering research must begin by locating the responsible group or groups." Or, "to answer the question 'normal or revolutionary?' one must first ask, 'for whom?" And again, "It is therefore, with respect to groups like these that the question 'normal' or 'revolutionary?' should be asked."[24]

Sociological studies have suggested that by 1951, when radio stars were optically identified, the research area of radio astronomy was established in principle. Furthermore the "various radio astronomy groups around the world, by then, knew clearly both what their experimental strategies should be, and what professional scientific audience was waiting for their results."[25] There was, therefore, a (paradigmatic) structure within which the discovery of pulsars was achieved.

Pulsar research, therefore, in the first instance is radio astronomy, and, indeed its *accoucheurs,* Hewish et al., were unambiguously known as radio astronomers. Smith, one of the early workers in the field, so designated them; he furthermore reveals that he "moved from radio to optical astronomy where observing possibilities in the pulsar field are very limited".[26] The subspeciality of x-ray astronomy also developed interest in pulsars as these objects were discovered to radiate not only at radio and optical frequencies but at x-ray ones as well. Therefore pulsar research now spans various astronomies. It has also stimulated work by condensed-matter physicists. Thus, considerable theoretical work has been done on the solid state of the interior of neutron stars and the crystalline state of the surface. According to the concluding remarks in a book arising from a symposium on the *Physics of Dense Matter,* "the majority of people presenting papers are not members of the IAU [International Astronomical Union]. A large number of the participants are physicists who have entered the field very recently because of their interest in the properties of neutron stars."[27] The point is that some workers in the field of pulsar research, which initially was the province of astronomers, came from other already established areas. In a sense therefore the milieu changed as the research activity progressed. The milieu responsible for the initial work and the establishment of the problem was clearly Radio Astronomy. However, this group was enlarged, as the subject developed; we assert that this expanded milieu clearly represents astrophysics or, more precisely, high-energy astrophysics: this distinction is underwritten by the origin of pulsars from supernovae, and by their association with cosmic rays.

The questions concerning the initial reception of the "discovery" of neutron stars would be addressed to Radio Astronomers. Questions concerning the continuing research effort might be more properly addressed to (High-Energy) Astrophysicists. Whether Kuhn envisaged a shifting milieu is another matter, but in a developing field it is not surprising that the milieu changed over the early years.

It is possible as well as instructive to justify further the attempt at isolating research groups by exemplifying the operational efficiency of such an undertaking. In particular, we shall be able to see how relevant it is to say that some scientific activity is normal for one group and not for another. Cosmology has been in an uncertain state for many years: big-bang, steady-state theories and their variants have been in competition during these years without any of them becoming established.[28] We might even say that cosmology is a set of theories in search of a paradigm. Now, cosmology conceivably might make use of the results of pulsar research to support any one theory, and to refute the others. It can do this without involving radio astronomy in any of the attendant in-fighting, so that at the same time that radio astronomy is executed in a normal setting, through the puzzle-solving skill of its experts (as we shall show), cosmology could be behaving in a crisis-ridden way. This kind of analysis and discrimination gives us a fuller appreciation of how (Kuhnian) normal and revolutionary science might differ from each other, and it thereby supports the distinction itself.

Having identified the research group we shall want to establish if possible, the dominanting paradigm under which it works, for, as we have remarked, Kuhn distinguishes between normal and revolutionary science and defines the former as operating under the unturbulent reign of a single paradigm. Furthermore, he assures us that this concept "points to the central philosophical aspect of my book."[29] Kuhn uses paradigm to mean both a disciplinary matrix *and* a particular element of this matrix—paradigm as shared example. In the first usage he would include symbolic generalizations that are incorporated by laws and theories, metaphysical components like belief in particular models, and values like the necessity for quantitative and accurate predictions; and secondly, and most important, paradigm as shared example. The latter use is, therefore, simply the most important element in the more global connotation, namely a constellation of group commitments.

In the postscript mentioned previously Kuhn agrees that it is problematic to have paradigm meaning two different things. He agrees to use disciplinary matrix for the first (global) meaning of paradigm. He then isolates an element of this matrix, shared example, as the central element of his book. It is this element that he would name paradigm except that "paradigm" has assumed a life of its own. He therefore agrees to label it as "exemplars."

Now, in connection with the first characterization of the Kuhnian paradigm, it is not difficult to establish that radio astronomers have group commitments and that these commitments are sufficiently distinctive to identify the group—that, in other words, the members operate within a disciplinary matrix. This group has flexible standards and requirements of accuracy. The ability to identify radio sources with optical ones depends crucially on the very accurate position-finding methods developed at Cambridge. Yet the need and ability of radio astronomers to know, for example, the surface temperature and composition of a neutron star does not match the corresponding need and ability of, say, metallurgists. They subscribe, in fact, like geophysicists and of necessity, to quite a lot of order-of-magnitude calculations. Both extremes of accuracy help to identify the groups. They are identified at one extreme by a Cambridge astronomer who reveals the extent to which instrumental accuracy was at the core of their commitments by remarking that "Whether the radio sources . . . were going to be supernovae, distant galaxies . . . or intelligent civilisations, one would want to be able to measure them better."[30] Apart from their flexible requirements on accuracy, radio astronomers or high-energy astrophysicists are observationally rather than experimentally oriented. This is necessarily so and distinguishes them, in a way that excludes even the geophysicist.

As to metaphysical elements of the matrix, radio astronomers subscribe to a hierachical structure of the universe, which is irrelevant to the physicist as solid-state electronics expert. Even more fundamentally, radio astronomers believe that the physics at the time and place of, say, Andromeda's radio emission is the same as that on earth now. (Crystallographers, for example, do not need this belief.) They, more than most other physicists, have a historical sense.

Equally fundamental but more crucial are the group commitments to what Kuhn calls symbolic generalizations. They are fundamental because they are not questioned; they are (more) crucial because it is through them that the group's activities bear fruits. Such generalizations encompass Einstein's field equations $R_{\mu\nu} - \frac{1}{2} g_{\mu\nu} R = 8\pi G T_{\mu\nu}$; Maxwell's electromagnetic equations, for example, $\nabla \times \vec{B} = 4\pi J$; and reaction equations like $p + e^- \rightarrow n + \nu$. Such generalizations merge with paradigm as shared example or exemplar. Thus the ability of radio astronomers to handle the physics and mathematics of a rotating magnetic dipole or of interfering electromagnetic waves goes continuously back to the undergraduate level. As far as laws and theories are concerned we can see from previous sections, what theoretical framework guided the research that led up to the discovery of pulsars. In this way general relativity, electromagnetic theory, and statistical and nuclear physics, for example, controlled the initial thrust of pulsar research. In a more direct way we can see also that in the actual execution of pulsar research during the 1960s and 1970s these same theories held uncontested sway and even the existence of various competing models of pulsar radiation does not undermine the validity of these arguments.

It is not only the shared example of setting up Einstein's general equation and solving it for various conditions that would have been the common experience of all the professional members of the group; but there are shared experimental techniques as well, probably not quite so common especially to those trained in the theoretical aspects of the discipline. The ability to use receivers and recorders and to analyze their charts would be among the background knowledge of this group. Whether theoretical or experimental work is involved, members of the group have acquired thereby tacit knowledge, and it is the nature of this tacit knowledge that defines the group.

It needs to be said that there is nothing unique about high-energy astrophysicists being interested in Maxwell's electrodynamic laws. Communication specialists share this interest. Nor are they the only group with flexible standards of accuracy. Geologists provide another example. It is the *totality* of their commitments that defines them.

So with respect to shared examples also high-energy astrophysicists, as a group, share in the widest and most sophisticated way the solving of Einstein's relativity equations, Maxwell's electrodynamic ones, *and* the construction and use of radio receivers. Pulsar researchers were the only ones who had seen worked examples; had themselves been involved in similar exercises related to the application of Einstein's general relativity equations—namely, the white dwarf solution; had explored the mathematical solutions of electromagnetic propagation across the universe; *and* were, at the same time, familiar with the radio receiver, and its capabilities and limitations. They and no other group were able to detect and interpret those first pulsar signals.

The preceding narrative does not prove but it more than indicates that there are areas in which we might define a paradigm for pulsar researchers, and it goes a little way to establishing what that paradigm might be. Thus, without exhausting the list of all possibilities, we can say that pulsar research is "paradigmatic," that is, it exists in a fairly well-defined disciplinary matrix, with its practitioners sharing the same ontological and epistemological commitments. Furthermore, they share in a deep, professional way what Kuhn calls exemplars.

That pulsar research has not initiated any mutation of values seems to be important support for this research as "paradigmatic" according to Kuhn. Indeed the existence and stability of the paradigm are reflected in the ease of assimilation of the observational and theoretical results. However, this was partially guaranteed by the prediction of neutron stars, the growing acceptance of their possibility, and by their frequent use as explanatory models. As Kuhn maintains, there are discoveries of facts and invention of theories, but the "discoveries" predicted by (paradigmatic) theories are more appropriately called detections. They are intimations of the normality of the science; they also indicate that scientists were trying to sort out *how* things could be fitted together rather than *why* they were as they seemed.

Pulsar Research as Nonrevolutionary but Cumulative

This section sets out to introduce other criteria for the normality of pulsar research. Pulsar research had revolutionary antecedents—special and general relativity, for example—but these changes of world view had taken place about two to three decades before the conception of the neutron star in 1934. By the time neutron star research had begun, scientists were operating within the disciplinary matrix bounded in part by relativity theories insofar as the release of nuclear energy could be understood from the special theory, and collapsing stars from the general theory. These theories had become part of the establishment: Tolman[31] could produce in one paper particular solutions of Einstein's field equations, which in the following paper Oppenheimer and Volkoff simply referred to and manipulated.[32] The point is that even if one goes back three decades before the neutron-star detection, one finds the supporting theories to have been well established. Pulsar research was not ushered in by any revolutionary theories or concepts.

From the opposite perspective we can see that pulsar research itself was not revolutionary. For general relativity, far from being undermined by pulsar research, was thought capable of being supported by it, and this possibility came out of a growing understanding of pulsars. We find a similar situation existing in relation to nuclear, statistical, and electromagnetic theories. Thorne and Ipser, and Bertotti et al. simply ruled out vibrating neutron stars on the basis of inadequate binding energy, and showed that instability of the pulsating configuration would result.[33] When Bertotti et al., Eastlund, and Komesaroff, for example, attacked the problems of pulsars they employed standard electromagnetic theory without any apologies.[34] They were not testing these established theories nor attempting to refute them; they accepted them as part of the constellation of their group commitment.[35]

We do not hold any brief for the future, and it is certainly possible that genuine anomalies might arise in connection with pulsars that will lead to crisis, revolution, and new physics. It is certain, however, that there were no revolutionary theories or other elements ushering in the birth of the neutron star or pulsar.

Although the historical sketch in earlier sections set the background for this very thesis, the development of radio astronomical techniques might appear to challenge this (continuity) thesis directly. First, it is very unlikely that pulsars would have been discovered without these techniques. Second, and to the point, the development of the important techniques took place *after* Zwicky and Oppenheimer et al. had done their work. One might therefore be inclined to see these instrumental developments as crucial and therefore perhaps as revolutionary. However, we can discount this apparent breach of our position on the continuity of pulsar research by noticing that the developments were narrowly instrumental. We must remember that whereas Galileo's telescope involved more than instrumental novelty, revolution even, in that it helped to change the metaphysical commitment regarding the perfection of celestial bodies and the baseness of earthy ones, radio astronomical techniques involved no such metaphysical enlightenment or conversion. The astronomical advance can be seen as a natural development of what had previously existed. Nor did the experiments set out to, or did in fact have to, adjudicate between rivals.

Another possible breach of our position is that researchers of the 1930s were not yet radio astronomers. It would thus appear that the group did not remain intact over the early period of our concern to which we have already adverted. Radio Astronomy began, indeed, only in a trivial way in 1931 with the observation of radio emission from the galaxy by Jansky;[36] it was the technical developments by Hey and Southwell,[37] among others, which were a by-product of the war effort, that led to its virtual inception more than ten years later in 1942. Its real establishment was achieved about ten years later as we earlier remarked. Radio Astronomy names a subject and thereby seems to name a group that we admit did not exist. The answer to this dilemma is provided by Kuhn. He denies that because a subject matter is named, a community is ipso facto named; he would deny the converse also. Thus a physics community existed in the eighteenth century although the subject matter did not exist linguistically. To repeat, a paradigm governs a group of practitioners, not a subject matter. Thus, although radio astronomers, so-called, did not exist an appropriate group did. We can therefore be satisfied that a continuous line may be drawn from the prediction of pulsars by Zwicky and Baade in 1934 to the work on pulsars more than thirty years later, during which time there was cumulative development.[38]

Indeed Bethe was building on Gamow, Landau on Zwicky and Chandrasekhar, and Oppenheimer et al. on all of them. Furthermore, Hoyle's theory of nucleosynthesis rested upon the foundations of Bethe's original conceptions not less than his "Relativistic Astrophysics" had direct ancestry in Oppenheimer et al. except that Hoyle et al. improved upon them in the greater thoroughness of their solutions. All this work led straight to the certainty of the existence of neutron stars with their predicted properties and associations: pulsars, identified as neutron stars, are the results of the "continuing . . . search for Astrophysical applications of general relativity."[39] This is what pulsar research has bequeathed us; and the resolution of the uncertainty about the mechanism of pulsar radiation is unlikely to affect the certainty of the star's existence.

Of course, by the early 1960s radio astronomers had gone beyond the stage of accepting the mere possibility of the existence of neutron stars. They were actively engaged in exploiting these stars for the explanation of previously known or recently discovered phenomena. From the initial detection of neutron stars, therefore, although there might have been a sense of excitement, there was no real surprise or any deep sense of anomaly. The detection of a neutron star (or black hole) would be merely helping to complete the sophisticated astrophysical picture that was being fashioned, and all of this supports one of Kuhn's criteria for normal science. It also describes a methodology concerned with discovering *how* scientists were trying to fit the pieces of a pattern together.

Pulsar Research as a Puzzle-Solving Activity

The absence of competition among established theories, values, and other components of the pulsar paradigm supports our contention that pulsar research is normal science. At the same time the existence of puzzle-solving activity as the basis of this research, as we shall show, reinforces this point as it demonstrates that pulsar research satisfies another major criterion of normal science according to Kuhn.

As Kuhn expands on the concept of normal science, he argues that: "Bringing a normal research problem to a conclusion is achieving the anticipated in a new way, and it requires the solution of all sorts of complex . . . mathematical puzzles."[40] Although the problem's outcome can be anticipated, the problem remains fascinating because the way to achieve its solution remains very much in doubt; and we also see clearly that such an activity responds to "how" and not to "why." Operating within such a framework is operating within a paradigm, and Kuhn asserts that "radio telescopes are only the most recent examples of the lengths to which research workers will go if a paradigm assures them that the facts they seek are important."[41] It is not necessary to stress that this pointed statement, relevant to our discussion but made many years before the detection of neutron stars, disposes of any lingering doubt about fitting the apparently revolutionary radiotelescope development into a relatively smooth and normal development of pulsar research. It supports also our reference to the puzzle-solving activity as being both theoretical and experimental.

We have remarked on the multitude of theoretical models introduced to explain radiation from rotating neutron stars. These models exist within the framework of established and accepted theories. What they represent are simply examples of scientists making use of their knowledge of tensors and expertise in the manipulation of Einstein's field equations of general relativity (Chiu and Occhionero); and a similar command of the intricacies of Maxwell's electrodynamic equations (Sturrock, Eastlund).[42] What Komesaroff, for example, did was to refer to and start from some electromagnetic result that was to be found in a standard textbook (Jackson's *Classical Electrodynamics*). He was behaving as a normal scientist, for Kuhn tells us that normal science "means research firmly based upon one or more past scientific achievements . . . such achievements are recounted . . . by science textbooks . . . which expound the body of accepted theory, illustrate many or all of its successful applications with exemplary observations and experiments."[43] Assuredly the theoretical models were all tests of some sort, but they were, more fundamentally, attempts to use familiar tools to solve a new but not abnormal problem.

The experimental side of the enterprise is also consistent with Kuhn's views of normal science. Normal science generally has narrow expectations and any "project whose outcome does not fall in that narrow . . . range is usually just a research failure."[44]

Thus Michelson-Morley experiments, performed within the pre-relativistic paradigm, was a failure for it was expected to show a fringe shift and it did not. Michelson himself regarded it as a failure. However, within the relativistic, Einstenian paradigm it was a tremendous success. Kuhn himself recalls another example, Coulomb's work on the attraction of charged spheres: "Coulomb and his contemporaries . . . possessed this . . . paradigm . . . that, when applied to the problem of attraction, yielded the same expectations . . . it is . . . why that result surprised no one."[45] This is exactly the point made earlier about radio telescopes used in the detection of neutron stars.

Kuhn has a further word, indirectly, on the nature of pulsar research. First, Kuhn refers to the dictionary definitions of puzzle that are illustrated by *jigsaw* puzzles, and then goes on to isolate the characteristics that these share with the problems of normal science. Further, in considering a situation similar to that of the proliferation of models for the radiation mechanism, he noted that "Many research problems . . . take this form. Tests of this sort are standard component of what I have elsewhere labelled 'normal science'. . . . In no usual sense . . . are such tests directed to current theory."[46] They simply form a puzzle-solving activity. These arguments are matched by the observation in *Nature* that in pulsar research the activity will involve "Astronomers . . . fitting pulsars into the *jigsaw* of stellar evolution."[47] We particularly note that the term *jigsaw*. We can think of the radiation models, therefore, in Kuhn's words, as "manipulation of theory." It is therefore clear, and worth emphasizing, both that pulsar research is not only a normal puzzle-solving activity, but is *seen* to be such, and that such an activity points to a methodology described as *how* rather than *why*. Pulsar research is a particularization of the general arguments put forward in the previous chapter.

Conclusion: Pulsar Research *Is* Normal Science

The nature of the scientific activity in pulsar research has been found to support Kuhn's theory of science. This support comes from the detailed background picture drawn in the early sections to show us the continuous and mainly traditional paths that led up to actual pulsar research. There were no revolutionary elements detected in its immediate prehistory, in its conceptualization, or in its execution. Indeed, the research carried out may be best described, essentially, as the work of experts rather than of geniuses. It was effected by the mobilization of puzzle-solving techniques, existing within a disciplinary matrix, and there were no competing paradigmatic theories. All these characteristics are to be found in Kuhn's depiction of normal science.

Of course, this is not the only theory of scientific practice. Popper and his followers provide formidable alternatives, but we are really mainly concerned, in this essay, in using Kuhn's theory that posits both normal and revolutionary science, not in arguing for it. However, we must pay brief attention to other methodologies. Popper himself has made a qualified admission of the existence of normal science, but his own theory does not incorporate it. This, we feel, is one of its inadequacies. Popper's theory is concerned with progress and growth in science, it is true, but he unnecessarily narrows his view of the whole of science to include, as Kuhn sees it, "only its occasional revolutionary parts."[48] More pointedly, Popper himself dislikes "normal" science, and "normal" scientists; the one because it is a danger to science, the other because they must have been badly taught. This is, of course, logically lame support for Popper in his confrontation with Kuhn, and we need not take these dislikes seriously in this context. All that we shall say is that normal science is not uncritical but rather acritical science: it is not mindless activity but only the reflection of no pressing need to question the foundations of the group's values, techniques, or theories. The emergence of positive anomalies will be time enough.[49]

Philosophers of science have not given very much support to Kuhn's theory although some scientists have done so. One philosopher, who supports at least one aspect of it is Pickering, who has made a study of a contemporary and still unresolved problem—the theory and "observation" of quarks. In this work he makes a study of the paradigm within which a group of workers continue their search for quarks. He supports Kuhn's effort and his particular articulation, and he laments the obscuring of Kuhn's original analysis "in the contemporary theory-oriented philosophical debate over incommensurability."[50] What Pickering says in detail is that "the problem posed for analysis [by quark research] is one of elucidating the relation between the theoretical communities and experimental practices sustained within communities of scientists."[51] Unlike most historians and philosophers, Pickering thinks that Kuhn has shown an awareness of this relation. What his study on quarks shows is that "scientific communities tend to reject data that conflict with group commitments and, obversely, to adjust their experimental techniques and methods to tune in on phenomena consistent with their commitments."[52] Whatever one's judgment of this activity, it clearly sustains a key characteristic of Kuhn's normal science.

It is highly suggestive that it is by analyzing major research activities like pulsar and quark research that support can be gained for the Kuhnian view of the scientific enterprise. I would suggest that it is not only in physics that such support will be found; there is a *prima facie* case that an analysis of DNA research will reveal similar practices.

We round out this section by referring to someone else who, despite criticisms, has agreed with major aspects of Kuhn's theory. This is Margaret Masterman whose expressed sentiments we share. It is true of the arguments as it is of hers that "far from querying the existence of Kuhn's 'normal' science, I . . . assume it." It is also true that "The present [essay] is written more from a scientific point of view than from a philosophic one." It is accepted also "That there is normal science—and that it is . . . the outstanding, the crashingly obvious fact which confronts and hits any philosophers of science who set out, in a practical or technological manner, to do any actual scientific research."[53] It is hoped that the foregoing pages have helped to support that contention.

Finally we note that, even after Kuhn's thesis is (broadly) accepted, we still need to recall the methodological framework within which these last two chapters have been written. Pulsar research as presented in this chapter is intended to exemplify a particular kind of categorical question that science always asks, namely *how* questions. In the context of our three categorical questions *what* and *why* had already been answered to a large extent by Zwicky and Baade on the one hand, and by Einstein and Oppenheimer et al. on the other. Methodologically, only *how* questions were left to be answered and scientists could and should have taken the path to the answers of these questions. We have seen that they did ask such questions, although there was some element of inadvertence in the actual "discovery" of pulsars. However, the reasons for this "failure" is not for us but perhaps for sociologists to judge. Indeed, Edge has reported that from "time to time, I have detected a sense of regret among radio astronomers that the process of discovery has been so accidental

. . . and I heard a Nobellist remark that perhaps we can look forward to radio astronomy becoming more scientific in the future!" Even so Edge finally sidesteps our issue by asserting that "The aim of sociology is to explain and understand not to evaluate and judge; to explicate, not to arbitrate."[54] The matter, perhaps, need only be noted and not pursued for, we would suggest, it was a human rather than a methodological "failure."

PART V

The Scientific Nature of Science

9

From *Scientia* to Science

Radio astronomers regretted that the process of discovery has
been so accidental . . . and . . . a remark was heard that perhaps we can look forward to radio astronomy becoming more
scientific.

—D. O. Edge, *Proceedings of the International Conference
on History of Physics*

Introduction

The "scientific nature of science" is not a paradox to be resolved
but a truism to be pondered. This compact description reflects
the development of the study of nature or natural philosophy out
of the logical cocoon of Aristotle's *scientia* into the enlightened
activity of scholars like Newton and Kelvin and Fermi. After
making "digressions" through Chapters 3 to 8, we are really returning to complete the arguments of Chapters 1 and 2 by showing how our present view of that body of knowledge called science, as concept and as activity, is a transformation from a
philosophical orientation in which science was *scientia* to the
modern framework in which mathematics and experiment are
the main buttresses.

We recall that "science," *scientia,* or *episteme* then denoted
"knowledge" but had the connotation of a special kind of knowledge: knowledge of things and their principles or causes. We
recall also that "science" now means such a structured and
definable activity that excludes music and metaphysics. It is
revealing to look at "science" from this (linguistic) point of view
and see it as always having been regarded as a special kind of
knowledge, which strengthens our links with the past and deep-
ens our understanding of the general cultural impact of science.

At the same time, therefore, this chapter sets out to show not
that the *special* status of "science" has changed but only that

199

which was and has always been special.[1] Original texts and articles among other sources will be examined to see how the changes took place philosophically, methodologically and linguistically. The period extends from 1605 to 1840,[2] from Francis Bacon's *Advancement of Learning* to Whewell's *Philosophy of the Inductive Sciences,* and the titles themselves give some indication of the changes as we move from "Learning," which Bacon himself translated as *Scientiae,* to "Sciences," which were defined by Whewell who also coined the word "scientist."

We ended our examination of prevolutionary science substantially with Roger Bacon who died toward the end of the thirteenth century, although we made various observations and comments on authors like Ockham, Chaucer, and Sir Philip Sidney who spanned the fourteenth to the sixteenth centuries. So there is essentially no gap as we pick up the story of Chapter 1 with Francis Bacon who was born in 1561.

It may be wondered whether Francis Bacon's works provide the next important methodological landmark in the development of science, especially when we reflect on what is known about giants like Kepler and Galileo. But after Grosseteste, Francis Bacon achieved the most extensive and detailed articulation of scientific methodology. It was so because his criticism went to the (logical) root of Aristotle's methodology, and he explicitly argued for one of the two significant changes that characterized the Scientific Revolution, and thus helped to launch the experimental tradition. That he did not go the whole way explicitly to include mathematization is offset by the repercussions his works created in France through Voltaire, and in Britain through the Royal Society. Whatever defects we now think that his method had—whether, for example, he was talking about observation rather than experimentation—it was taken very seriously and developed. The case gains support, I think, in that whereas others like Ockham, Buridan, and Kepler, for example, had made significant criticisms of particular Aristotelian views, it is Bacon who both perceived the need for, and proposed a comprehensive change to, the foundation of Aristotle's system. So the significance of Francis Bacon is that he not only had innovative ideas, but he produced a detailed plan of action.

Now Aristotle separated the various areas of knowledge, all parts of philosophy, through their individual axioms. This approach had failed to draw significant distinctions and, in any case, had become sterile. This is exactly where Bacon attacked: he condemned Aristotle's method as causing us "to revolve for ever

in a circle, mak[ing] only some slight and contemptible progress." In particular, "The art of logic has tended more to confirm
Errors, than to disclose Truth." He thought that the sterility was
evident in "our present Sciences (which) are nothing more than
peculiar Arrangements of matters already discovered, and not
methods for discovery of plans for new operations." He felt that
"An instauration must be made from the very foundations." This
is very stirring and revolutionary stuff; but we add that Bacon,
too, was concerned with the traditional "certainty" and "demonstration," asking him who wishes "to know to a Certainty and
Demonstration, as a true son of Science . . . (to) join with us."
Bacon's significance is further enhanced by his expounding that
"Experience is by far the best Demonstration."[3] He is casting
aside Aristotle's Logic, which involved philosophers merely with
differentiating axioms, and with achieving geometrically tidy
conclusions.

What Bacon intends to do is set out clearly in *Novum
Organum* where he argues against the prejudice and inadequacy
of the Aristotelian program with its emphasis on the mind. He is
going to join mind and hand in an equal endeavour. As he says in
Aphorisms 2: "Neither the naked hand nor the understanding
left to itself can effect much. It is by instruments and help that
the work is done, which are as much wanted for the understanding as for the hand." Similar sentiments are expressed in *Aphorisms* 94 and 96.[4]

More specifically, Bacon, while retaining the deductive scheme
of Aristotle, removes the emphasis on the syllogism, and he
emphasizes the inductive part only now related to experiments.
For he indeed believed that the best natural philosophy ascended
from experiments to causes and descended from causes to new
experiments. It is clear that Bacon was not discarding the whole
of the Aristotelian programme[5] but modifying it. However, his
modifications were profound.

To a large extent such considerations, as we have now introduced, deal with Bacon's methodological and philosophical positions, and as yet ignore his linguistic usage. We tracked down
both philosophical and linguistic usage in the earlier Bacon, and
we shall be equally concerned with those in the later Bacon.
However, we shall approach his linguistic usage from a consideration of his classification of knowledge.

Bacon's Classification of Knowledge

Ideally what we are looking for is the approximately contemporaneous particularization of *"scientia"* and "science"—that is, an

equivalent semantics of both words, which go beyond Roger Bacon on the one hand and certainly beyond Chaucer and Sir Philip Sidney on the other. Although Francis Bacon is especially useful in this exercise because he wrote both in Latin and in English, and actually translated one of his books from English into Latin, there is no guarantee that we shall find in him what we are looking for.

These particularizations might be expected to reveal modern connotations of the word or the concept. We shall find that Bacon's linguistic usage is not quite in phase with his methodological program, especially its modernist orientation, and to begin with, it certainly seems that Bacon is stuck with one of the traditional usages. Most spectacularly, the title "Advancement of Learning," in which book he does consider the whole range of academic and practical disciplines, is translated twenty years later, under his supervision, as *De Augmentis Scientiarum*. Furthermore, when he refers to the three knowledges (divine, natural, and human philosophy), he is quite positively using "knowledges" to describe the traditional *scientiae*.

Other examples, which on the contrary suggest a modern usage, are somewhat ambivalent. For example, in *Novum Organum* in which Bacon is discussing sciences as we know them, he refers to the superficiality of most scientific discoveries, believing that "Quae adhunc inventa sunt in scientiis hujusmodi . . . notioribus vulgaribus fere subjaceant; ut vero ad interiora et remotiora naturae penetretur," which may be translated as "The discoveries which have hitherto been made in the sciences lie close to . . . the surface, in order to penetrate the recesses of nature."[6]

From the context the unqualified use of *scientiis* could have meant "knowledges." The later reference to *naturae* suggests, however, the unvarnished *scientiis* could have referred to the natural sciences, which implies a certain particularity. We shall need a more careful examination of Bacon's usages, which we can obtain from a study of his classificatory scheme.

Aristotle and Gundisalvo were classifers of knowledge and so was Francis Bacon. When he settles down in "The Advancement of Learning" to discuss natural philosophy, he asserts that "these be the two parts of natural philosophy . . . *natural science* and *natural prudence*." These are further described as theoretical and practical so that *"natural science,"* despite appearances, still seems to fall short of the modern connotation, restricted as it is to a theoretical part only. This impression is further intensified

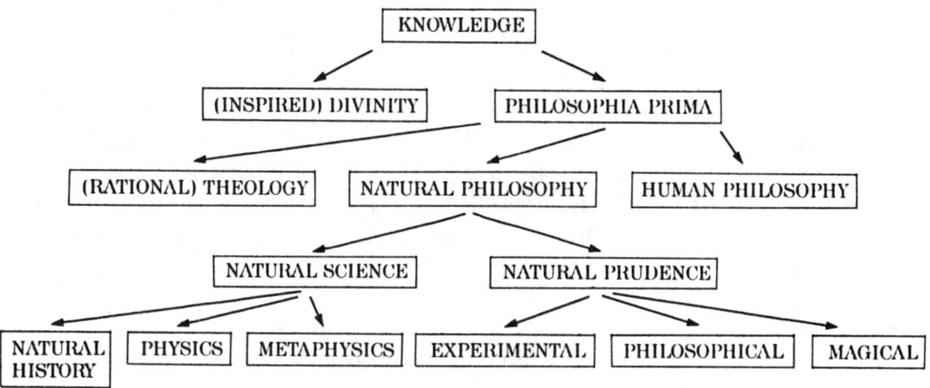

when he joins *Physique* and *Metaphysique* (!) as representing
the two parts of natural science, for *physique* was for Bacon,
science, the subject dealing generally with matter and motion.
Thus, *physique* would consider the cause of the melting of wax
whereas *metaphysique* would consider the nature of heat.[7]
We find on further investigation at the structural and philosophical levels that Bacon has modern notions about natural
science in that there is a theoretical as well as an experimental
part. The linguistic usage, however, is out of phase in that such
science is called natural philosophy, the theoretical part being
labeled natural science and the practical part natural prudence.
Thus, we have the classification shown in the figure.
We shall ask questions of this classification at the detailed level
(What are the sciences?), at the structural level (What is science?),
and at the metascientific level (What is the role of science?) We
might thus obtain a more accurate view of the relationship of
Bacon's "science" to modern "science."
The modern structure of Bacon's science may be immediately
appreciated when we notice that "natural science" encompasses

all of what we would now call science: the factual histories (natural history), matter and motion (physics), and the unavoidable metaphysical components of scientific theories. If "magical" seems medieval and retrograde, this is only linguistic deception for it meant for him merely natural wisdom. For the other notions of magic he had nothing but sarcastic dismissal, describing "these sciences which hold so much of imagination and belief, as this degenerate Natural Magic and the like, that in their propositions the description of the mean is ever more monstrous than the pretence or end." And he elaborated, with literary relish: "Of this kind of learning the fable of Ixion was a figure who deigned to enjoy Juno . . . ; and instead of her had copulation with a cloud, of which mixture were begotten centaurs and chimeras. So whosoever shall entertain high and vaporous imaginations, instead of a laborious and sober inquiry of truth, shall beget hopes and beliefs of strange and impossible shapes."[8]

Bacon furthermore links Alchemy and Astrology with the Magical, so that clearly a lot of what we would now regard as nonscience failed to qualify for Bacon also: for "it is as far differing in truth of nature from such a knowledge as we require." And what was that knowledge? Bacon was in no doubt that "he that knoweth well the natures of weight, of colour, of pliant and fragile, in respect of the hammer, of volatile in respect of the fire and the rest"[9] will be much more efficacious in his experiments. Thus in detail we are satisifed that Bacon's sciences was, to a large extent, our sciences.

At the general level of structure we find the experimental not as well integrated with the theoretical as we would like. However, both are not only placed on the same level but Bacon himself acknowledges that both theoretical and experimental work are related, that "as there is an intercourse between causes and effects, so as both these knowledges . . . have a great connection between themselves."[10] He saw the theoretical part as deductive and the experimental part as inductive, but both as essential for the prosecution of natural science. He separated them therefore for reasons that he said were not philosophical but literary.

Furthermore, although Bacon might seem to underrate mathematics, as is often remarked, this is not so. It is oddly entered in his classification as part of (Theoretical) natural science, where however it is to operate as a tool. Bacon himself believed that the greatest satisfaction for man's desire to embrace "the spacious liberty of generalities" lay in Mathematics to the pure branch of which "those sciences . . . which handle quantity determinate" belonged. Mathematics was rather oddly placed within Bacon's classification but it was not underrated. He believed that "the placing of this science . . . [was] not much material."[11] It was not underrated, and others have found no evidence for his underrating of the subject.

In considering finally the role that Bacon saw his new science as playing, we shall find ourselves, quite naturally, investigating that other important aspect of Bacon's writings, his linguistic usage. By this we mean the semantics that can be associated with his use of *"scientia"* and "science." Bacon had a view of science as an orderly activity governed by rules and laws; not static rules for the performance of a task or the execution of a blueprint, but a dynamic one for the creation of new knowledge. Indeed, he wished that natural philosophy would concentrate on the "discovery of forms," and we are assured that "when I speak of forms, I mean nothing more than those laws and determinations of absolute actuality which govern and constitute any simple nature, as heat, light, weight, in every kind of matter and subject that is susceptible to them."[12] A law-directed modern science is envisaged. Thus, he laments that "no one successfully investigates the nature of things in the thing itself; the inquiry must be enlarged to become more general." Bacon believed that workers often failed to do this because of bad attitudes, for "even when they seek to educe some science or theory from their experiments, they nevertheless almost always turn aside with overhasty . . . eagerness to practice, . . . for the sake of the uses and fruits of the practice."[13]

Clearly in "educing some science or theory from their experiments," Bacon's workers were doing what modern scientists do; and we note especially that the existence of the activity supports the view of a modern, organic relationship between theory and experiment.

But what was this "science" that Bacon's workers were educing? What does the classification and the metascientific positions tell us about Bacon's linguistic usage? The classification places natural and human philosophy on the same plane, and rational theology also. Does "science" cover all of these? It is true that his method was developed to be applied to natural as well as to human sciences but, in fact, the latter are only briefly and sketchily dealt with. Indeed, Bacon introduces the issue, nervously it seems, when he wonders whether "it may also be asked whether I speak of natural philosophy only, or whether I mean that the other sciences, logic, ethics and politics, should be carried on by this method."[14] Whatever his intentions were, the fact is that natural philosophy is the predominant science to which his method is applied; and against his actual naming of ethics and politics as sciences as we have the emergence of natural philosophy as the most dominant science, perhaps the prototypical science. The role of Bacon's science was modern.

Yet we are stuck with the linguistic residue. The naming of activities or concepts is not a mere lexicographic exercise; indeed, Bacon was working in an established cultural milieu, and we can understand the recalcitrance of the word "science" if we understand this milieu. Bacon, in introducing *natural science,* thinks that in dividing natural philosophy into two parts he would "do best [to] allow a division . . . in familiar and scholastical terms . . . the *inquisition of causes,* and the *production of effects; speculative,* and *operative; natural science* and *natural prudence.*"[15] Thus "natural science" is to be linked with "speculative" and "inquisition of causes" as well as "scholastical." Furthermore "scholastical" takes us back to the medievalists' *scientia,* and "familiar" reminds us that it was the current usage even in Bacon's time.

We know that "scientia" is linked with the medievalists' "inquisition of causes" by simple reference to a text[16] on scholastic philosophy. There a distinction is made between speculative philosophy and operative philosophy (philosophy of life), the former being identified with *"the natural knowledge of all things through their last causes."* It is then noted that this knowledge is not mere knowledge but *scientia,* defined by the scholastics themselves as "systematized knowledge . . . especially knowledge through causes." Such a citation supports the thesis of our Chapter 1 as well as points out for us the semantical luggage that Francis Bacon had acquired: his "natural science" still retained vestiges of the old *scientia.*

Nevertheless, whatever were the received meanings of *"scientia"* and "science," we cannot ignore the impact of Bacon's two books, and especially the new feeling for and meaning of science that they conveyed. Indeed, at the most elementary level, *The Advancement of Learning* and *Novum Organum* are predominantly concerned with (our) natural science. At a deeper level "science" and *"scientia"* are employed in these books, without qualification, to apply to natural science. This kind of emphasis, for Bacon and for his readers, must have meant a growing and deepening association of physics and biology with "science" with its new and modern structure.

It must be admitted that one is predisposed to link Bacon's natural science with ours, despite the linguistic counterevidence, because it is acknowledged that Bacon's science did indeed differ greatly from Aristotle's—did in fact have this new and modern structure. We can therefore accept that a modern reading is embedded in the linguistic entity *"scientia"* or "science."

Such ambivalence is bound to exist at the frontiers of knowledge, and however significant Bacon was as a modern precursor, it must be admitted that he did not get it all in place in those epochal books. No one (except Einstein?) ever does. Galileo still referred to circular motion as inertial motion, although he clearly grasped the essentials of the concept and had a set of demonstrations to verify the correctness of that concept. Even Newton, in his law of motion, referred to *vis insita* as if, in the medievalist tradition, we were to think of inertial motion as fueled by some incorporeal, internal force. Language rarely moves in step with concept, and its habitual usage tends to distort even those novel and brilliant concepts that change the face of science.[17]

Institutional Developments

Bacon was a great influence in the developing concept and practice of science, and the fruits of that influence can be savored through the men and institutions that flourished so abundantly in the seventeenth century. There were many men working in the field of science but Newton was the central figure, the one who conquered and scaled the "Everest" of the scientific endeavor. He both climbed the inductive slopes using the newly charted experimental method as exhibited in his *Opticks;* and enjoyed exhilarating descents down the deductive slopes as revealed in his *Principia.* Kuhn himself refers to the classical sciences and the Baconian sciences—the one theoretical and relatively independent of experiments, the other experimental without much recourse to theory. Examples of these are Astronomy and Electricity, respectively, as they existed in the seventeenth century. Kuhn's view is that Newton was the first man to bridge[18] these two kinds of sciences, in his *Principia* and in his *Opticks.*

For a century and a half after Newton's death, a large cohort attempted to find shorter, more beautiful and more fruitful paths, starting from the ones this collossus had marked out. This large-scale activity introduced sophistication into the scientific synthesis first begun by Newton, and it is from the pinnacle of his achievements that we shall view the transmission of Bacon's influences.

One of the major factors in the development and spread of Newtonian science was the scientific societies that sprung up in Italy, France, and England between 1603 and 1666. Bacon was a specific reference in the establishment of the Royal Society which originated partly from "something . . . offered about a design of founding a college for the promoting of physico-mathematical, experimental learning."[19] Whatever vagueness remained in the connotations of Bacon's "science," it is clear that through Bacon these institutions, with such a blueprint, were bound to impress an individuality on their area of study that would affect a wide cross-section of people. For, although the metropolitan societies might cater to the *cognoscenti,* within decades of the formation of these societies other institutions were developing in England, in other urban areas, and in the provinces also, where the general public provided the main audience.

We need to view the emergence of these institutions not only as influential on succeeding generations but, to some extent, as a set of legitimizing acts. The institutions arose out of the realization that there was a body of men with common, fairly well-defined interests—in other words, "scientists." Thus, the mere existence of these institutions spoke of a particularity that science had acquired.[20]

Equally important in this development of the concept and practice of science was the accompanying professionalization. Bacon had already argued for a career in science, akin to religion, in its selflessness and dedication, and he spoke of scientists as of a priesthood. He also prophetically, divided "natural philosophy . . . to make two professions . . . of natural philosophers."[21] It was no surprise, therefore, that the institutions that he inspired produced men who assumed roles at least equal to those assumed by doctors, lawyers, jewelers, or priests. Newton was the high priest. We have made reference to Preserved Smith in an earlier chapter where he describes that period when "With an intensity of conviction almost religious . . . a chosen band of disciples set out to educate the public in the (Newtonian) principles which, they believed, would prove efficacious . . . in explaining the operations of nature." Their hopes were sanguine because Newton himself had given explanations of centuries-old mysteries of the world and beyond, until then within the prerogative of God.

Professionals were indeed emerging, and it seemed to people like Fontenelle that *a new kind of man* (my italics) was being produced that would be associated almost entirely with the practice of science. Fontenelle praised one academician for his "character [which] was as simple as his superiority of mind could require. I have already given this same praise to so many persons in this academy that one would believe the merit pertains rather more to our sciences than to our savants."[22] Fontenelle is commenting upon the emergence of the identifiable scientist as well as upon the cultural impact of the related science. He was indirectly making reference to the double conception of the profession and the professor.

Meanwhile, other changes related to this double conception were taking place, more insidiously perhaps. Thus Brockliss has analyzed the way in which the new physics (science) was replacing the Aristotelian one in the universities in France, particularly in Paris.[23] Brockliss determined that by 1690, with as much as twenty professors teaching physics, all the Aristotelian influence had disappeared, and Cartesianism had replaced it. "Modern" science was becoming widespread, and Brockliss argues that "the acceptance of a new set of explanatory theories was accompanied by a change in the attitude to the epistemological foundations of natural philosophy."[24] The new science was in place even if it was not consistently named.

"Modern" science became even more widespread from the popularization of Cartesianism and of Newtonian natural philosophy by people like Fontenelle and Voltaire, as we saw in Chapter 5. The natural philosophy of the eighteenth century was clearly acquiring a popularity that could only arise from a distinct perception of it; certainly the "conflict" voiced by Voltaire could only have arisen from such a distinction. Furthermore, in the eighteenth and nineteenth centuries interest in science spread from the salons and universities to the general lecture rooms. Concerning Davy's lectures at the Royal Institution in 1801 it was reported that "men of the first rank, . . . the practical and the theoretical . . . women of fashion . . . the young all crowded . . . eagerly . . . the lecture theatre."[25] Now, persons from the highest to the lowest ranks had always been interested in technology. For the Mediterranean peoples "Astronomy" was technology that allowed them to chart their trade journeys from port to port. Up to the fifteenth century it was certainly very popular with the so-called average man. When the compass was discovered it superseded "Astronomy" as a technological instrument. Astronomy always had a theoretical base and developed theoretically, but it was left to the philosophers Galileo and Kepler. On the other hand, the interest among the many auditors at Davy's lectures was in the principles as well as the practices of natural philosophy—that is, in a new science, distinct from *scientia*.

Throughout the late eighteenth and early nineteenth centuries people came to a more direct apprehension of natural philosophy as a separate and self-sufficient area of study in other ways. Despite the expression of some views that the steam engine was just a craft invention, Kerker has argued that in the seventeenth and eighteenth centuries scientists were "vitally interested in the practical utility of science, and many of their problems were oriented towards subjects useful for technical development."[26] Perhaps the earliest engineers Papin, Newcomen, and Savery engaged in raw technology, but this is doubtful. Watt also may not have used Black's discovery of latent heat, directly, in his model engine after he was first introduced to a Newcomen engine; but Newcomen's apparatus was introduced not as a machine but as a laboratory instrument. However, he used scientific principles in his analysis of the engine. Certainly he "studied the physical processes involved and where necessary, carried out independent experiments . . . [which] led . . . to the rediscovery of the latent heat of steam, to the relation between pressure and boiling temperature."[27] The final products of engines and trains were clearly not the result of God or philosophy or pure craft but of some new science. By the end of the eighteenth century we had not only the institutionalizing of science and its profession by a varied body of men, but the tangible results in which the natural philosophers and the people could share. These various changes resulting from an increasingly sophisticated methodology were inevitably accompanied by linguistic ones also.

Science and Scientists

One way of catching the emergent common usage of science is to rummage through the books and especially the journals of the time. There we should discover how some of the leading thinkers referred to what we would all call natural science and scientists.

We note that journals were a direct result of the formation of those institutions inspired by Bacon's work.

In the mid-eighteenth century natural philosophy was mainly physics with its mathematical framework experimental method, and Newtonian authority. These things defined a species of study, and in 1781 Kant was denying that "Chemistry was a science since it lacked a deductive mathematical system."[28] When, in 1787, he changes his mind after the work of Lavoisier and Priestly, it is clear that his conception of science was modern. Kant uttered his judgment on the basis that "cognition that can contain merely empirical certainty is only improperly called science"; for him "Natural science properly so called presupposes . . . laws . . . [and] in every special doctrine of nature only so much science proper can be found as there is mathematics".[29] One would be finnicky if one denied this as an acceptable and historically documented statement of the modern notion of "science" and of its structure and function.

The case, however, is made more persuasive by corroboration from an English source, and indeed, in the highly esteemed Edinburgh Review, we have much praise being bestowed on Dr. Black, the discoverer of latent heat, that his "acquirements were not merely those of a man of science";[30] they included music as well. Music is no longer one of the sciences, which it had been two hundred years earlier. We have the unambiguous reference both to an identifiable activity that men practice, and to the profession of a "man of Science," or the "scientist" who was still to be named. The same review comments, in 1804, not about Rumford the experimental philosopher but the "experimentalist [who has] . . . unquestionably rendered some service to science"; and quite significantly, the commentator does not think "very highly of Count Rumford's talents as a philosopher."[31] What is meant by a philosopher is clarified when we read that "His general remarks . . . are far from being . . . original, and serve to diminish . . . our respect for [his] philosophical powers . . . [he] should leave the . . . framing of hypotheses[32] to inferior men . . . whose errors . . . can have no detrimental consequences on science."[33] Twice we have science clearly distinguished as a profession and, when it is taken in the context of Rumford's work, it defines our modern science. The statement also defines the practitioner, and Rumford is seen as an experimentalist who sometimes employs the philosopher's stock-in-trade. Here we have a clear statement that acknowledges a separation between science and philosophy, and between scientist and philosopher. That study, which no longer centered itself on philosophy, would certainly cease being natural philosophy. It ceased being *scientia* and became science.

We finally obtain an early example not just of the use of "science" but of its (modern) definition from Thomas Thomson in 1812.[31] He distinguishes between the study of substances and the study of changes in substances, describing the first as *Natural History* and the second as Science. The distinction is more fully articulated when he cites Chemistry as the subject dealing with imperceptible motions and Physics as dealing with visible measurable motions. Thus, Science *is* natural science and no longer *Scientia.*

"Science" was not coined; it emerged; and it was only after its emergence that men of science as identifiable people could be named. The Edinburgh Review of 1804 refers to Black as a man of science, and in the other issue refers to Rumford as an experimentalist doing science. This buildings up to Whewell's coinage of "scientist" between 1834 and 1840. Whewell realized the "need . . . [for] a name to describe a cultivation of science in general."[35] He had discussed also "why some portions of knowledge may properly be called science" and indicated that the "sciences to be treated are those which are commonly known as the Physical sciences."[36] This is a clear-eyed, articulated recognition of the existence of the profession and the professor. In fact, it is more than this. It casually betrays the extent to which the modern notion of science had already been accepted that the divisions "physical" and "biological" had appeared. Biology had been coined in 1802 by Lamarck in France and Treviranus in Germany to mark the distinctive scientific study of the phenomena of living matter. The modern concept of science had indeed established its growing acceptance and sophistication in this division.

The further distinction between natural and other sciences was not long in coming, although there is still much controversy about the accuracy of the labeling, over a century after Comte first introduced his Social Physics and the social sciences in the 1830s. This long controversy reveals to what extent the modern concept of science is that of natural science; it must be a rare student indeed who is in any doubt that the unqualified use of science means anything but natural science. The length of the controversy reveals also the extent to which the other "sciences" aspire to the condition of natural science.

It was the eventual success of Darwin's theory of evolution that set the final seal upon the study of natural phenomena as an autonomous activity, detached from traditional philosophy and metaphysics, although like Newton's physical science it had introduced its own metaphysics. Darwin's theory was not immediately successful; however, its impact was very great. This was due largely to the very personal nature of its concerns, and its influence was very widespread, affecting sociology, politics, religion, and philosophy. In the aftermath of Darwin's publications, the nineteenth century had one of its most serious intellectual convulsions centreing on the clash between Literature and Science and the extended debates between Arnold and Huxley, which we discussed in Chapter 5. All the debates were equally important as signs of the times but the Arnold-Huxley debate in addition has historical links with our immediate concern. As we will recall Huxley had observed that from 1850 there had been a battle between ancient and modern literature, and he saw science as a new element that had joined this battle. He asserted very vigorously that science could provide at least as good a basis for education as literature. Arnold was aware of Tyndall as "a brilliant and distinguished votary of the natural sciences" but he emphatically denied that natural science could form the main part of an education. This open warfare was simply the culmination of the nervous fears expressed by Voltaire almost a century and a half before. In the mid-nineteenth century the probability, even, of battle was some indication that a new and identifiable enemy had arisen, at least of equal stature with Literature. The simple fact of the battle was the unambiguous sign that Science had indeed arrived.

Conclusion: The New Science

The new science had arrived with the transformation of *scientia,* and it might appear that it was at tidy evolution from the clearsightedness and daring of Francis Bacon, through the genius of Newton and the massive talents of his French pupils like Lagrange and Laplace, to the actual verbalizing in the Edinburgh Review and in Whewell's works. It arrived, however, via a more complex route than this nevertheless correct outline suggests.

The ultimate seal of "science" was stamped on Natural Philosophy by its recourse to laws and theories, mathematically framed and essentially predictive, with its predictions subject, in principle to experimental disproof. Whether science describes or explains, coordinates or facilitates is a matter for what science does rather than what it is, which is our present concern. What it is can be better understood by looking more carefully at the path taken, and the way in which modern science developed out of two distinct strains. The major strain in science had been its rationalism from Aristotle and Plato to the medievalists, and laws and theories were part of this rationalism, however rhetorical they might have been. Thus Aristotle believed in the law that appeared in its medieval formulation as "Omne quod movetur ab alio movetur." Kepler is regarded as having introduced the first *mathematical* laws into science, and together with Descartes, whose Analytical Geometry could be applied to Kepler's ellipses, they both made indelible marks on the rational side of Natural Philosophy.

The other strain in science has been observational/experimental, and in the modern era it was philosophically launched by Francis Bacon. Galileo, with qualifications,[37] may be thought to have been the first to make a lasting impact on experimental science through his inclined planes, as this apparatus demonstrated the accuracy of his law of falling bodies. It was Newton, as we saw, (and here we follow Kuhn[38]) who was the first *scientist* in the sense that he first displayed the ingredients of science as defined. Thus his *Principia* contained the mathematical side of the rationalism of science, in which we have both his laws of motion and his gravitational theory; and his *Opticks* in which we detect genuine experimentation leading to a theory of colors. We note in passing that his gravitational theory approaches the scientific ideal or definition even more closely through its historical prediction of the existence of Neptune, detected in 1846. Science, however, is not made by one scientist who, in any case, regarded himself as a natural philosopher.

Thus we have the two strains of science, elsewhere referred to as those of "thinking" and "doing." Now "natural philosophy" carries with it the tone of philosophy, which has always been regarded as concerned with higher things of the mind. Indeed, the historical fact of the very gradual development of science is a partial reflection of the long-lasting distinctions between thinking and doing; and of the relative disparagement of the latter, which is exemplified by the admonition of St. Augustine "Go not out of doors. Return into thyself. In the inner man dwells truth."[39] Thus the rationalist strain of science maintained its early prominence.

Compounding and further exemplifying this prejudice was the fact that in the (late) seventeenth and eighteenth centuries, despite the institutional support for the activities both rational and experimental, there developed the distinction between generally empirical sciences like electricity and heat without theoretical base, and theoretical sciences like astronomy with only (limited) observational base. Science awaited not only the integration of theory and experiment but the removal of the more superficial distinction between theoretical and experimental sciences—in fact between thinking and doing.

The rationalist tradition had been the longest and the deepest held so that even when the study of nature was being methodologically separated from philosophy, the practitioners remained "natural philosophers," and the practice "natural philosophy." They were by then practicing "science," but this word was still appropriated by meanings of long standing and definitely related to the rationalist tradition. Thus, as we saw, for hundreds of years *scientia* and *philosophia* were interchangeable. And even Francis Bacon, who tried to remove the taint from "doing," retained the word "science" for the mental and theoretical part of natural philosophy or what we would call science. But this was not simply an English vice. It was not necessarily a vice for at the same time the empirical side was being developed with great skill, in England, by scholars who with their theoretical counterparts were yet called natural philosophers. Thus, the naming of all those workers as natural philosophers superficially united them, and the distinction between thinking and doing was thinly obscured. Furthermore, it was not only an English problem for in France the theoretical side of Newton's theories was brilliantly extended by scholars who were called "physiciens" or "physicists," whose activity was called *physique,* consistent with the rationalist connotation in the Baconian classification. Experimental work was referred to as *physique experimentale.* We are therefore led to ask what was the relationship between "natural philosophy" and *"physique,"* between the "natural philosopher" and the "physicien"?[40]

It is clear that "scientists" of both England and France were involved in the same activity, even though against the dominating empirical work of Boyle, Hooke, Newton, Bradley, and Black we had the equally dominating theoretical work of d'Alembert, Lagrange, Fourier, and Laplace. For such a difference is less stark than it appears, seeing that at the same time there was the theoretical work of Newton as well as the experimental work of Coulomb. The difference in emphasis was not a fundamental one; probably no more than a matter of a difference in opportunities that arose at that time. There was no serious difference of view as to what science was, for when serious communication began between both countries *"physique"* was rendered as "natural philosophy" and vice versa. This was an important equivalence because the traditional natural philosophers were now being indirectly but distinctly named, and such a change reflected new perceptions that embraced the unification of thinking and doing, a modern expression of science at least at the linguistic level.

At the same time the other theoretical-experimental distinctions, already referred to, were being substantially removed. The empirical, Baconian sciences were acquiring a theoretical base through du Fay, Fourier, and Black, for example; and the classical, theoretical sciences were being subjected to rigorous experimental investigations through the work of Cavendish, for example. The exchange of the knowledge and practice of physics between France and England helped to remove all distinctions, linguistic and methodological.

By 1840 Whewell had definitely settled the matter as far as we are concerned, for the linguistic seal almost invariably comes after the conceptual, practical, and institutional ones. Thus in the same way that in 1803 the Edinburgh Review identified a "scientist" without inventing the name, so we have seen that "science" had been identified in many ways before being named in 1796, in 1803, and definitely in 1840. We dimly recognized its existence in the work of Francis Bacon in 1620, and rather more clearly in Kant in 1786, but we would hazard the estimate that by about 1700 science had generally acquired its modern connotations, and had been thought about in the modern way.[41] The rapid communication that began taking place between France and England helped to obscure and then to remove all residual distinctions due to either the tradition or prejudice of centuries. The final naming of the activity was a relatively minor episode in this complicated journey, even though it arrived at the confluence of major developments like (institutional) support for scientific activity; the (sociological) recognition of the distinct benefits that these activities would bring; and the sophisticated mathematization of Baconian (as well as classical) sciences.

That which we call science had been perceived as one, whether the "doing" and the "thinking" had been (spuriously) unified under *"physique"* or "natural philosophy." Those whom we call scientists, who theorized, experimented, and generalized, had also been perceived as one whether they were called "physiciens" or "natural philosophers." The perception was the fundamental transformation, the naming was simply putting a seal upon a *fait accompli.*

10

Classical Duality in Modern Physics

> For classical physics has only gradually shaped the notions of particle and wave to logical contraries or opposites . . . [from 1900] physics was . . . divided into two: physics of matter based on the concept of corpuscles . . . , and secondly radiation physics based on the concept of wave propagation in a hypothetical continuous medium.
>
> —Max Jammer and Louis de Broglie,
> *Conceptual Development of*
> *Quantum Mechanics/Nobel Lectures*

Introduction

In keeping with the pattern established for this book we come finally to a particular example of science, as science is now perceived. The choice of the example avoids the (necessary) generalization of much scientific journalism; I hope it avoids also the denseness of much academic reporting on science. The rationale for the actual choice, seen to be located in the classroom, is that we thereby complete the list of arenas considered in which physics has been actively pursued over the centuries. Such a choice also reinforces the image of the subject as being alive.

The suggestion is made that some of the conceptual difficulties experienced by students in their introduction to wave mechanics could be eased by the pedagogical use of a historical approach involving the optical-mechanical analogy. This analogy appeals to work done by Hamilton, Fermat, and Maupertuis, who worked in classical mechanics and optics.[1] A unified approach will be used in this chapter both in the classical domain toward ray optics and particle trajectories, and in the quantum domain toward wave optics and wave mechanics. The purpose is to show the structural, aesthetic necessity for wave mechanics to exist as a generalization of classical mechanics. This approach should remove the daze of novelty that many students experience in their first brush with the theory. The counterintuitiveness of the theory may thereby become manageable.

At the very least the approach will bring this essay full circle to a satisfying close by showing the *historical* continuity of the concept of wave-particle duality, which was treated in Part I as part of the *philosophical* nature of science. By revealing also how important were the *aesthetic* criteria in this development, this chapter will make links with our earlier discussions in Part II. Finally, embedded in the discussion of this chapter are the fundamental why-questions that were part of the *methodological* considerations of Part IV.

It will be seen therefore that the particular choice that this chapter embraces is appropriate for impressing upon its audience how closely and perenially related are the natures of science.

Background Material

In using a popular approach to the teaching of wave mechanics, it is normal to give a historical sketch. In this connection the usual line is to show how the Rayleigh-Jeans and Wien approaches[2] led to incomplete correspondence with the empirical data, and how Planck was able to interpolate some data points to arrive at his famous equation.[3] We think that this background ought to be filled out in three ways. It is first necessary to indicate that several physicists were engaged on the problem of finding the spectral distribution of cavity radiation, and thus Planck's work fitted, as Kuhn would have it, into a "normal" problem thought by many to be worth solving. It is then necessary to explain that it was the unique and absolute aspect of the black-body spectrum that attracted so many first-rate physicists: its absoluteness made it special. Both these related facts put Planck's work[4] in much better perspective. It is then important to say that although initially Planck only "teased out" the correct formula, he actually derived it from fundamental principles soon after. It can then be more readily accepted that Planck's work received the Nobel Prize.

It is also very important to make the point that, although Planck's theory implied a discrete rather than the (classically accepted) continuous *exchange* of energy, he himself conceived of the quantization of the energy of material oscillators *only*, not taking the apparently logical step to consider the quantization of the cavity radiation with which they were interacting. Einstein did this shortly after. These are necessary and interesting historical points to make, but neither is the one with which this chapter will be actually concerned—namely, wave-particle duality.

Not until deBroglie, in 1923, suggested the duality of material particles did wave-particle duality (again) become a live issue. However, we must be reminded[5] that wave-particle duality was historically antecedent to the establishment of wave mechanics, and we shall emphasize this point in our treatment of the optical-mechanical analogy, which will be seen to be grounded in this duality. The analogy thus becomes historically relevant and acquires additional merit as an introduction to wave mechanics properly so called.

The Optical-Mechanical Analogy

The analogy was first formulated by Hamilton[6] whose aim it was to give to optics the same beauty, power and harmony with which mechanics had been endowed by Lagrange. Hamilton made his own powerful contribution to mechanics and, as Jammer reminds us, it was done during his investigations on optics. The analogy is in fact part of a more comprehensive scheme that we wish to employ. It is represented in diagramatic form in Figure 1.

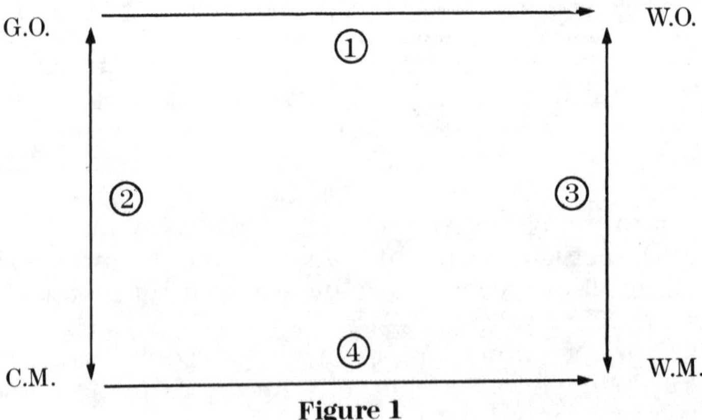

Figure 1

Figure 1 is constructed on the following arguments.

1. It is well known, and will not be argued, that geometrical optics is an approximation of wave optics, and so the arrow points toward the more comprehensive description. This is link (1).

2. Furthermore, there is an analogy between geometrical optics, with its rays, and classical particle mechanics with its trajectories. This will be analyzed and presented here as the optical-mechanical analogy. This is link (2).

3. Now Einstein suggested the particle (or mechanical) aspect of light (waves). DeBroglie postulated the (wave) aspect of (mechanical) particles or his "pilot" waves on the basis of symmetry. Both of these suggest a wave (optical)-mechanical analogy on the right, with the implication of a wave mechanics. This is link (3).

4. However, it is necessary to go beyond this rhetorical plausibility and discover whether our classical mechanics is an approximation of the implied wave mechanics. The working out of the deBroglie and Einstein dualities takes us there. In particular, there are certain implications of the optical-mechanical analogy on the left, and if we can show that, with a couple quantum concepts (which do not necessarily imply a wave mechanics!),[7] there is a similar implication of the optical-mechanical analogy on the right, then the postulation specifically of a *wave* mechanics is inescapable. We can then close the loop through the link (4), with the arrow head pointing towards wave mechanics.

We shall now pursue the path set out above.

Establishing the Analogy (i)[8]

GEOMETRICAL OPTICS

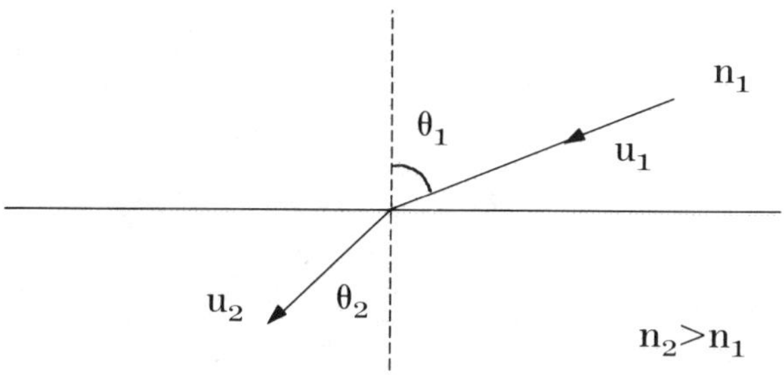

Figure 2

Geometrical optics deals with rays. In Figure 2 such a ray travels with velocity u_1 in medium (1) and is refracted at the interface of mediums (1) and (2) and then travels with velocity u_2 in the second. Snell's law for refraction at the interface gives

$$n_1 \sin\theta_1 \;=\; n_2 \sin\theta_2$$

or
$$\sin\theta_1/\sin\theta_2 \;=\; u_1/u_2$$

because
$$n \;=\; 1/u \tag{1}$$

where n represents the refractive index of a medium.
Here we have the refractive index being inversely proportional to the wave (ray) velocity.

PARTICLE MECHANICS

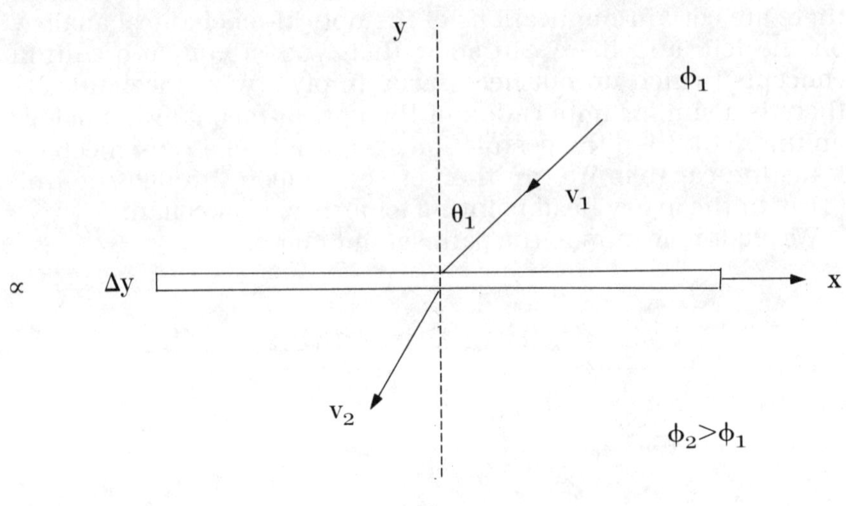

Figure 3

In Figure 3 a (charged) particle travels in one region of uniform electrical potential, ϕ_1, before entering a small region over which the potential changes, the change being $\Delta\phi$. It emerges into another region where the potential is again uniform, but now ϕ_2.
Now, force $= -\Delta\phi/\Delta y$. Thus the negative y- component of the

momentum of the particle will increase across the discontinuity; the x-component will remain unchanged on the assumption that the electric field is in the y-direction only. Thus $\tan\theta$ must decrease and refraction take place towards the normal. In the diagram v_1 is the particle velocity on entry into the region of discontinuity; the force increases it to v_2 with which it leaves the region.

Because generally the kinetic energy $\frac{1}{2}\,m\,v^2 = e\phi$ where e is electronic charge

$$v_1{}^2/v_2{}^2 \;=\; \phi_1/\phi_2$$

that is, $\qquad v_1/v_2 \;=\; (\phi_1/\phi_2)^{\frac{1}{2}} \equiv n_1/\,n_2$

It is valid to identify the potential with an equivalent refractive index, namely

$$\phi^{\frac{1}{2}} \equiv n$$

because the "bending" of the particle is indeed effected by the potential.

Since $\partial\phi/\partial x = 0$, the horizontal component of the particle momentum remains constant as was mentioned previously.

$$\text{So} \qquad p_1\sin\theta_1 \;=\; p_2\sin\theta_2$$

$$\text{that is,} \quad m\,v_1\sin\theta_1 \;=\; m\,v_2\sin\theta_2$$

$$\text{that is,} \qquad n_1\sin\theta_1 \;=\; n_2\sin\theta_2$$

where $p \equiv mv$ is the linear momentum.

This equivalent Snell's law establishes an analogy at the level of "laws": the optical-mechanical analogy.

Of course, here,

$$n \propto v \tag{2}$$

This is an awkward relationship, for if we wish to establish an analogy between the optical ray and the particle trajectory, then the relationship between refractive index and velocity should be the same in both cases. This awkwardness resides in the simple observation that whereas v increases across the mechanical discontinuity, u decreases across the optical discontinuity. This "embarrassment" will only be sorted out at the end.

FERMAT'S PRINCIPLE

However, we can pursue the analogy to a deeper level. Snell's law is simply derivable from a more fundamental optical principle, namely Fermat's principle of least time for the propagation of light from point A to point B as indicated in Figure 4. The

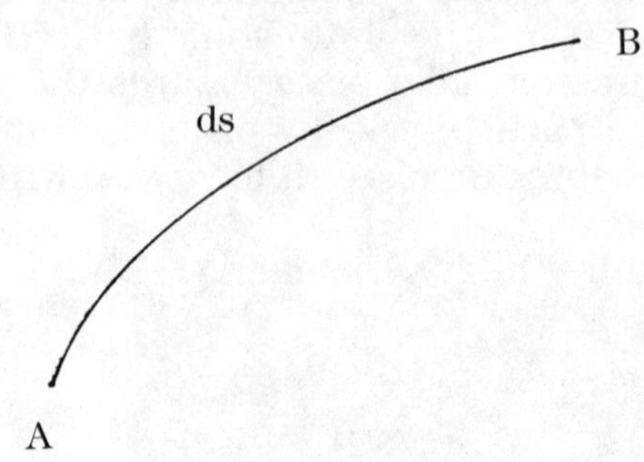

Figure 4

principle is saying that the actual path that light follows between two points A and B is the path that takes the least time.[9]

Because the time over the path is a minimum its variation, δ, must be zero—that is, $\delta \int \frac{ds}{u} = 0$ (where u is the wave velocity over *ds*).

This may be written as

$$\delta \int_{A}^{B} n_o ds = 0 \text{ and } n_o \equiv \text{optical refractive index} \qquad (3)$$

MAUPERTUIS'S PRINCIPLE

In classical mechanics there is the corresponding Maupertuis's principle:

$\delta \int 2\,T\,dt = 0$ where T is the kinetic energy of a mass point for which $T + V = E = \text{constant}$, where V is potential and E total energy of the mass point traveling from A to B in Figure 4.

Because $\qquad\qquad$ v dt = ds

and $\qquad\qquad\qquad$ $T = 1/2m\,v^2$

we get $\qquad\quad$ $\delta \int m\,v\,ds = 0$

which is the Principle of Least Action where Action, *A*, is defined as: $dA = p\,dq$—that is, $A = \int p\,dq$ where *p* is generalized momentum and *q* is generalized coordinate. It is of interest to record that Maupertuis discovered his principle of mechanics while working on the theory of light. The condition may be written alternatively as

$$\delta \int_{B}^{A} n_m \, ds \;=\; 0 \qquad \text{because } n_m \propto v. \tag{4a}$$

where $n_m \equiv$ mechanical refractive index.

We therefore have identical fundamental principles for optics and mechanics $\delta \int n \, ds = 0$. Although there is a discrepancy in connection with the relationship between refractive index and velocity for the optical and mechanical cases, we might suspect that it can be explained away because of the similarity, for the two modes, of the more fundamental principles.

The problem is that, if there is to be a valid optical-mechanical analogy, we should have $u \propto v$, whereas through equations (1) and (2) it is more like

$$u \propto 1/v \tag{4b}$$

or $u = k/v$. However, u is phase-velocity; and it is the group velocity, u_g, with which we should be concerned as it is *the* velocity with which optical energy is transferred from A to B. We note that v is the particle velocity with which energy is transferred from A to B mechanically. Thus if we found $u_g = k/u$, we would get $u_g = v$ and our analogy would be complete.

Establishing the Analogy (ii)

We need first to argue for an optical-mechanical analogy on the right. We analyzed laws and principles that were similar, on the left, for optics and mechanics; but we shall not be able to do exactly the same on the other side. However, from an a priori position one could argue that an analogy or similarity could be posited generally, if (optical) waves were seen as, and could be established to behave like, (mechanical) particles. This is one duality established by Einstein and Compton.[10] On the basis of symmetry and completeness one would have to establish, also, that (mechanical) particles could behave like waves. This is the second of two dualities "established" by Davison and Germer, and which we assume as an "axiom" of the optical-mechanical analogy on the right.

This analogy, therefore, is supported by the

(i) use of symmetry as a heuristic principle;

(ii) conceptualizing of a (double) wave-particle duality. Thus if we take up the analogy for electromagnetic waves by accepting their duality, then $v_o = \frac{1}{h} E_m$, where o and m refer to optical and mechanical properties. If we pursue the mechanical (right) side of this duality, again relativistically, we get, because

$$E^2/c^2 = p^2 + m_o{}^2 c^4$$

and
$$E/c = p.$$

$$\lambda \equiv c/v = \frac{E/p}{E/h}$$

that is,
$$\lambda_{(o)} = h \cdot \frac{1}{p_m}$$

A consideration of the duality of *waves* therefore gives us

$$v_o = E_m/h$$
$$\lambda_o = h/p_m$$

These are the Einstein-deBroglie relations and express, partially, the optical-mechanical analogy on the right. If we use our heuristic principle, and duality is pressed independently for *particles* then we have, accompanying the particles, "pilot" waves with wave or phase velocity

$$u = \lambda v$$
$$= h/p \cdot E/h = E/p \tag{5}$$

from the duality relations noted previously.

The important point, however, is that if we take the analogy, seriously, and these "pilot" waves do represent the particles precisely,[12] then the implication of equation (4b) must follow also from the analogy on the right. We should be able to deduce $u = k/v$ where v is the particle velocity and u the corresponding wave velocity. Furthermore, and equally importantly we must find that the energy is transmitted with a velocity ($u_g =$) v. The first question, therefore, is whether the analogy on the right entails $u \propto 1/v$.

Let us then find a relationship between energy and momentum for *particles*.

For particles: $E = mc^2$ relativistically.
Also $p = m\,v$
$\therefore$ $E/p = c^2/v$ for particles, where v is particle velocity. $\tag{6}$

Now, a valid representation of particle by wave must mean, in the first place, that

$$E/p)_{\text{particle}} = E/p)_{\text{wave}}$$

that is,
$$c^2/v = u \tag{7}$$

Thus, the phase velocity for the "pilot" wave u is $\propto 1/v$, where v is the corresponding particle velocity. We recall that we found, for the analogy on the left when considering rays and trajectories, an identical relationship. We see that the optical-mechanical analogy on both sides of the equation

$$\text{G.O.} : \text{C.M.} \equiv \text{W.O.} : \text{W.M.}$$

entails the same result, namely

$$u = k/v.$$

The analogy may be regarded as explicitly established on the left hand whereas it is only implicitly established by the wave-particle duality on the right hand. However, we are willing to accept it on both sides because of the identical entailments. It becomes entirely acceptable if we can show $v = u_g$.

To begin with, we have generally, for the relationship between group and phase velocities,

$$u_g = u - \lambda \frac{du}{d\lambda} \tag{8}$$

If we make use of the following relationships, already introduced:

$$\lambda v = u$$

$$p = h/\lambda$$

$$E = h v$$

$$E^2 = p^2 c^2 + m_0^2 c^4$$

Then
$$\frac{du}{d\lambda} = \frac{d}{d\lambda} \cdot (\lambda \frac{E}{h}) = \frac{d}{d\lambda}\left[\frac{1}{h}(h^2 . c^2 + m_0^2 c^4 \lambda^2)^{1/2}\right] \text{ because } p = h/\lambda$$

$$\therefore \quad \frac{du}{d\lambda} = \frac{m_0^2 c^2 \lambda}{h(h^2 c^2 + m_0^2 c^4 \lambda^2)^{1/2}}$$

$$\therefore \quad u_g = \frac{E\lambda}{h} - \frac{m_0^2 c^4 \lambda^2}{h\lambda E}$$

$$= \frac{1}{h\lambda E}(E^2 \lambda^2 - m_0^2 c^4 \lambda^2)$$

$$= \frac{1}{\lambda h E}(h^2 c^2 + m_0^2 c^4 \lambda^2 - m_0^2 c^4 \lambda^2)$$

$$= h^2 c^2 / h^2 u = c^2 / u = v$$

Thus the analogy is established on both sides through identical entailments: $u = k/v$; and the apparent difficulty of this entailment is explained away by showing that $u_g = k/u = v$.

There must therefore be, we argue, a *wave* mechanics that more precisely describes the behavior of particles in the same way that wave optics more precisely explains the behavior of light.

Conclusion

If students find this presentation as persuasive as its neatness, logic, and plausibility demand, then they will move into the rigors of the course with at least one psychological block removed: they will have accepted the *existence* of a wave mechanics. Presumably most of what that acceptance implies will also be less forcefully resisted. Thus, operators, commutators, and Heisenberg's uncertainty principle; complex wave functions, eigenvalues and probability densities; and certainly wave-packets and duality will all be *accepted* even if not fully *understood,* because of the perception of the inevitability of their framework—that is, of wave mechanics.

This chapter achieves at the same time the aim consistent with the structure of the book in that it shows, in a particular although limited way, modern science in action. We went through a mathematical exericse in a (modern) way of which we had no glimpse in the earlier chapters. Furthermore the exercise was supported both by fundamental principles, themselves of a mathematical nature, and by experimental observations of a fundamental nature also. This was indeed a cameo of modern science.

It is to be hoped also that the fact that this chapter has helped us to encompass in a natural way, in one essay, the logic and physics of Aristotle as well as the abstruse argumentation of modern scientists like deBroglie and Schrödinger will not be lost on any reader of this essay. In this chapter we have the example of "duality" as a concept of scientific importance in the twentieth century, but it brings with it the "scientific" ideas of the fourth century B.C. as well as aesthetic ones. Through the concept of duality the theme of this book has been realized. However, we will not cajole any readers into accepting that science has profound

and perennial virtues; the most that we will hope for is that if
this exercise has any merit, then no one will put down the book
and yet remain empty-handed—that indeed each will take according to his inclination and his appetite.

Notes

Chapter 1. Science as *Scientia*

1. Sydney Ross, "The Study of a Word," *Annals of Science* 18 (1962): 65–85.
2. John Ruskin, *The Works of John Ruskin,* ed. E. T. Cook and Wedderburn (London: George Allen, 1908) 22:396n; 34:157n.
3. Brian Stock, *Myth and Science in the Twelfth Century* (Princeton: Princeton University Press, 1972); Tina Stiefel, "The Heresy of Science: A Twelfth Century Conceptual Revolution," *Isis* 68 (1977): 347–62; Ross, "The Story of a Word." See also Tullio Gregory, "La nouvelle idée de nature et de savoir scientifique au XXIe siècle," *Boston Studies* (Reidel, Holland) 26 (1975): 212; Richard McKeon, "The Organization of Sciences and the Relations of Cultures in the Twelfth and Thirteenth Centuries," *Boston Studies* 26 (1975): 151–92; and Edith Dudley Sylla, "Autonomous and Handmaiden Science: St. Thomas Aquinas and William of Ockham on the Physics of the Eucharist," *Boston Studies* 26 (1975): esp. 349–54.
4. The subsequent section on Grosseteste and Roger Bacon may be thought, in the context of this chapter's purpose, as somewhat self-indulgent, for its argument takes us beyond the establishment of science as *scientia*. To this extent it may be omitted on a casual reading. Yet its argument for a perceived change in Roger Bacon reinforces the case for the original status of science. The arguments also form a bridge between this chapter and Chapter 9.
5. Aristotle, *The Basic Works of Aristotle,* ed. Richard McKeon (New York: Random House, 1941).
6. Aristotle, *Metaphysics* 980.
7. Ibid., 1005b6.
8. Aristotle, *Posterior Analytics* 71b8–23.
9. Ibid., 72a9.
10. Ibid., 74b16.
11.

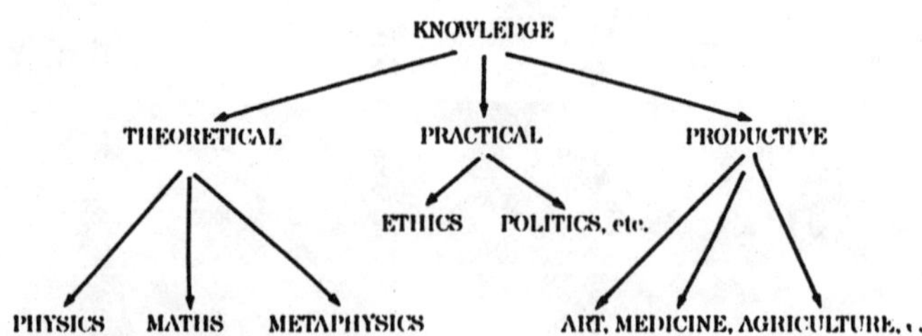

12. Aristotle, *Metaphysics* 981b30.
13. Aristotle, *Physics* 194b20.
14. Aristotle, *Posterior Analytics* 79a23.

15. Ludwig Edelstein, *Roots of Scientific Thought* (New York: Basic Books, 1957), 94–95.

16. Aristotle, *Posterior Analytics* 87a30ff.

17. Ibid., 72b20.

18. Aristotle, *Posterior Analytics* (London: Heinemann, 1960), 155, 157.

19. Aristotle goes on to say that "demonstrations are . . . universal and universals are imperceptible, [so] we . . . cannot obtain scientific knowledge by the art of perception. . . . if we were on the moon, and saw the earth shutting out the sun's light, we should not know the cause of the eclipse." This discussion gives us a concrete example of the application of Aristotle's principles to what we would call science or scientific knowledge.

20. Boethius, *Consolations of Philosophy* (London: G. Routledge and Sons, n.d.), 188.

21. Daniel D. McGarry, *The Metalogicon of John of Salisbury* (Berkeley: University of California Press, 1955), 214.

22. Ibid.

23. A. C. Crombie, *Robert Grosseteste and the Origin of Experimental Science* (Oxford: Clarendon Press, 1953), 36–37.

24. Karl R. Popper, *Conjectures and Refutations* (London: Routledge and Kegan Paul, 1969), 93–96.

25. Stock, *Myth and Science*, 31

26. Gregory, "La novelle idée de nature." Gregory believes that the major difference in twelfth century concepts was that before then divine causality played a predominant role in the explanation of physical phenomena. From that century, however, events were given a physical explanation according to a natural cause.

27. Stiefel, "The Heresy of Science," 348.

28. Ibid., 360.

29. Crombie, *Robert Grosseteste*, 52n.

30. Ibid., 59n.

31. Ibid., 54.

32. Roger Bacon, *Fr. Rogeri Bacon Opera* ed. J. S. Brewer (London: Longman, 1859), 33.

33. Roger Bacon, *Opus Maius* (New York: Russell and Russell, 1962), 2:583, 615.

34. Ibid. Bacon is clearly referring to something that is more than experimental. We believe that however much his "experiment" fell short of the modern notions, it was not just "experience."

35. Ibid., 587ff.

36. Ibid., 116.

37. Ibid.

38. Stewart C. Easton, *Roger Bacon* (Oxford: Blackwell, 1952), 131–43; 188–96.

39. A. G. Molland "Medieval Ideas of Scientific Progress," *Journal of the History of Ideas* 39 (October–December 1978): 562.

40. Ibid.

41. This general point gains some support from Bacon, *Opus Maius*, 79 who laments that "almost all the secrets of philosophy up to the present time lie hidden in foreign languages. For as in many instances only what is common and worthless has been translated." Furthermore it is in the use of "science" in early English that the notion, quite accurately, arises that it meant no more

than knowledge. "Science" did mean knowledge in its pristine Latin form but a sophisticated usage had developed that naturally eluded the grasp of the dependent language as we contended.

42. Walter Skeat, ed., *The Complete Works of Chaucer* (Oxford: Clarendon Press, 1912), 101.

43. Ibid., 499.

44. Ibid., 680.

45. Ibid., 396.

46. W. Peacock, ed., "The Voyages and Travels of Sir John Mandeville," in *English Prose,* chosen and arranged by W. Peacock (London: Oxford University Press, 1921), 1:22.

47. Ibid., 307.

Chapter 2. Classical Duality and the Nature of Light

1. Louis de Broglie, "The wave nature of the electron," in *Nobel Lectures; Physics 1922–1941* (Amsterdam: Elsevier Publishing Co., 1965), 245.

2. Aristotle, *Categories* 1b10

3. Ibid. It is necessary to state that in Aristotle "substance" was not equivalent to "matter." A substance was never just matter. "One kind of substance exists . . . as matter, another as form . . . while the third . . . is that which is composed of these two."

4. Aristotle, *On the Soul* 418b18.

5. Vasco Ronchi, *The Nature of Light* (London: Heinemann, 1970), 61.

6. S. Sambursky, *The Physical World of Late Antiquity* (London: Routledge and Kegan Paul, 1962), 7.

7. Edward Grant, ed., *A Sourcebook in Medieval Science* (Cambridge, Mass.: Harvard University Press, 1974), 379.

8. Ibid.

9. David C. Lindberg, "Alkindi's Critique of Euclid's Theory of Vision," *Isis* 62 (Winter 1971): 470–71.

10. Ronchi, *The Nature of Light,* 124–133

11. Aristotle, *On the Soul* 418b27.

12. Ibid.

13. Grant, *Sourcebook,* 418.

14. Bacon, *Opus Maius,* 489–90.

15. Johannes Kepler, *Paralipomena ad Vitellionem* (Frankfurt, 1604), 9. Cited in Ronchi, *Nature of Light,* 88–9.

16. Michael Faraday, *Experimental Researches in Chemistry and Physics* (London: Taylor, 1859), 480.

17. Michael Faraday, *Experimental Researches in Electricity* (New York: Dover, 1965), 1:79.

18. Faraday, *Experimental Researches in Chemistry and Physics,* 480.

19. Grant, *Sourcebook,* 230.

20. Aristotle, *Physics* 200b33.

21. Ibid.

22. Grant, *Sourcebook,* 229, n. 10.

23. It would probably be regarded as useful, at least, to eliminate all the

accidents except motion if we wish to establish the inevitability of the use of motion to describe light. Ockham's work provides a model where he argues about light that "such a thing is not substance, for it is neither matter nor form nor anything composed out of them. Nor is it a quality, as can be inductively proved by running through all the species of quality. Nor is it quantity . . . and so on." Even Ockham does not go tediously through the list, and we may be satisfied with his analysis. In any event the case for "motion" is taken up in a different way later in this chapter.

24. Aristotle, *Physics* 200b12.

25. A. C. Cotter, *ABC of Scholastic Philosophy* (Weston, Mass.: The Weston College Press, 1949), 370.

26. Ernst Cassirer, *The Problem of Knowledge, Philosophy, Science, and History since Hegel* (New Haven: Yale University Press, 1950), 87–88.

27. A. I. Sabra, *Theories of Light from Descartes to Newton* (London: Oldbourne Book Co. Ltd., 1967), 32–33.

28. We must note that Grosseteste seemed to support an accidental, wave description *as well as* a substantial one.

29. Bacon, *Opus Maius*, 489.

30. Ronchi, *The Nature of Light*, 130.

31. William Magie, ed., *A Sourcebook in Physics* (Cambridge, Mass.: Harvard University Press, 1969), 283.

32. I. B. Cohen and R. E. Schofield, eds., "Newton's Paper on Light and Colours," in *Isaac Newton's Papers and Letters on Natural Philosophy* (Cambridge, Mass.: Harvard University Press, 1958), 184.

33. Sambursky, *Late Antiquity*, 113.

34. Sir Isaac Newton, *Opticks* (New York: Dover Publications, 1952), 362.

35. In classical physics the accepted view was that *particle* and *wave* were mutually *exclusive* descriptions of natural phenomena. It is consistent with our thesis that the Medieval question, framed as *substantia aut accidens*, employed for "or" the exclusive *aut* and not the inclusive *vel*, as indeed Jammer points out.

36. We note, also, that Weisskopf believes that "the significance of quantum mechanics becomes clearer when we realise that only the wave-particle duality can give us those stable and regenerating patterns that form the basis of Nature." He was talking about the "morphic" character of quantum mechanics that, he argued, had "introduced the elements of form, shape and symmetry into physics." Furthermore, he believed that aspects of these forms, such as stability against change, were combined by Nature into the various structures that we find around us, and that duality played the significant role in producing them. See "Is physics human?" in *Physics Today* (June 1976): 26.

Chapter 3. Aesthetics in the History of Scientific Theories

1. G. Santayana, *The Sense of Beauty* (New York: Dover Publications, 1955), 15.

2. William Durant, *The Story of Philosophy* (New York: Simon and Schuster, 1953), 2.

3. Ernst Cassirer, *The Philosophy of the Enlightenment* (Princeton: Princeton University Press, 1951), 342.

4. Henri Pioncaré, *Science and Method* (New York: Dover Publications, 1952), 59.

5. Erwin Schrödinger, *Science, Theory and Man* (New York: Dover Publications, 30. Schrödinger writes also that he "need not . . . speak of the quality of the pleasures derived from pure knowledge; those who have experienced it will know that it contains a strong aesthetic element and is closely related to that derived from the contemplation of a work of art . . . [lying] beyond purely utilitarian activities." However, he does not entirely dismiss the view that the beauty of an object may originate in its utility; he believes that "if the criterion of usefulness be thoroughly carried out it will evolve its own type of beauty."

6. Lancelot Whyte, ed., *Aspects of Form* (Lund: Humphries, 1951), 3–5.

7. Lancelot Whyte, Accent on Form (New York: Harper and Row, 1954)

8. J. P. de Crousaz, *Traité du Beau* (Amsterdam: F. Honoré, 1715), chap. 7.

9. Harvey, for example, used quantitative arguments in analyzing the "circulation" of the blood.

10. C. F. von Weizsäcker, "Physics and Philosophy," in *The Physicists Conception of Nature,* ed. J. Mehra (Boston: Reidel Publishing Co., 1973), 737.

11. Kuhn includes music among the classical sciences as opposed to the Baconian sciences (which we discuss in Chapter 9). He was reluctant to include it together with astronomy until he read sufficiently deeply to discover how highly developed it was, and mathematically rigorous.

12. Tobias Dantzig, *Number* (New York: Free Press, 1967), 43.

13. Ibid.

14. Ibid.

15. Santayana, *Sense of Beauty,* 8.

16. J. L. Dreyer, *History of the Planetary Systems from Thales to Kepler* (Cambridge: Cambridge University Press, 1906), 179.

17. Santayana, *Sense of Beauty,* 164.

18. Arthur Koestler, *The Sleepwalkers* (New York: Grosset and Dunlap, 1963), 254.

19. G. D. Birkhoff, *Aesthetic Measure* (Cambridge, Mass.: Harvard University Press, 1933), chap. 1 and 2.

20. Crousaz, *Traité du Beau,* chap. 7. "When our different faculties are equally satisfied, it is unity which charms us in the midst of variety." (My translation)

21. Agnes Arber, *The Mind and the Eye* (Cambridge: Cambridge University Press, 1964), 92ff.

22. René Descartes, *The Philosophical Works* (Cambridge: Cambridge University Press, 1972), 1:263.

23. Cassirer, *The Philosophy of the Enlightenment,* 289.

24. A. Koyré, *Newtonian Studies* (Chicago: University of Chicago Press, 1965), 7.

25. Ibid.

26. Poincaré, *Science and Method,* 23.

27. J. L. Anderson, *Principles of Relativity Physics* (New York: Academic Press, 1967), 329.

28. Wolfgang Rindler, "Relativistic Cosmology," *Physics Today* (November 1967): 23.

29. D'Arcy Thompson, *On Growth and Form* (Cambridge: Cambridge University Press, 1966), 271, 285.

30. Poincaré, *Science and Method,* 60.

31. Jean Ullmo, "The Agreement between Mathematical and Physical Phenomena," in *The Critical Approach,* ed. M. Bunge (New York: Free Press, 1964), 356, 358.

Chapter 4. Formal and Connotative Aesthetic Elements in Physical Theories

1. This does not mean that we equate the scientist with the artist for it can be argued with Kuhn that scientists solve puzzles, and may use aesthetic criteria to achieve their goals. On the other hand artists create aesthetic objects, and the solution of puzzles may help them to achieve these goals.

2. Birkhoff, *Aesthetic Measure,* chap. 1

3. N. Campbell, *Physics: The Elements* (Cambridge: Cambridge University Press, 1920), chap. 8.

4. Mario Bunge, "The Complexity of Simplicity," *The Journal of Philosophy* 59 (March 1962): 113–35.

5. Ludwig Wittgenstein, *Lectures and Conversations on Aesthetics, Psychology and Religious Belief,* ed. C. Barrett (Berkeley: University of California Press, 1967), 13.

6. Koyré, *Newtonian Studies,* 165.

7. Albert Einstein, *Sidelights on Relativity* (London: Methuen, 1922).

8. Michael Polanyi, *Personal Knowledge* (London: Routledge and Kegan Paul, 1958), 145.

9. Birkhoff, *Aesthetic Measure,* chap. 2. The connotative response is mainly external but not entirely so because there must be some aspect of the theory on which this response feeds; and this aspect might have its own internal aesthetic.

10. Frank Ramsay, *The Foundations of Mathematics* (London: Routledge and Kegan Paul, 1931), 291.

11. Dantzig, *Number,* 87. Indeed Dantizig, taking the same attitude to the case, remarked (almost, one might think, as a retort to Condillac) that in so far as mathematization involves the use of symbols, a symbol has "a meaning which transcends the object symbolised: that is why *it is not a mere formality.*"

12. Nelson Goodman, "Recent Developments in the Theory of Simplicity," *Philosophical and Phenomenological Research* 19 (June 1959): 429. Much of the work of the last few decades has concentrated on the simplicity of hypotheses, involving quantification based on the number of assumptions and of parameters. The results have not been conclusive but, in any case, there are other aesthetic elements apart from simplicity, as we shall see, and they are not susceptible to quantification. The failure to quantify simplicity should not therefore be held against this element and, by extension, against the whole aesthetic enterprise.

13. Karl Popper, *Objective Knowledge* (Oxford: Clarendon Press, 1972), 197. It is the same kind of simplicity to which Popper refers when he says that "The 'depth' of a scientific theory seems to be most closely related to its simplicity and so to the wealth of its content."

14. There is occasionally a lack of sharpness between epistemological and

psychological simplicity. This is certainly related to Cassirer's view that an adequate explanation of the human mind is to be found in its evolution and that, therefore, psychology is designated as the foundation of epistemology.

15. C. S. Sharma, "The Role of Mathematics in Physics," *British Journal for the Philosophy of Science* 33 (September 1982): 275. Sharma has argued, in relation to general relativity, that it "has always used models which are intuitively rather difficult to grasp."

16. F. Rawlins, "Methodology and Aesthetic Content in Natural Science," *The Institute of Physics Bulletin* (September 1959): 205.

17. Jacob Bronowski, *Science and Human Values* (New York: Harper and Row, 1965), 19.

18. M. Scott, ed., *The Essays of Francis Bacon* (New York: Scribners' Sons, 1908), 198. Yet, we have Francis Bacon's famous saying that "There is no excellent beauty that hath not some strangeness in the proportion!" It is also to be noted that Harry Woolf's book about the aesthetician par excellence, Albert Einstein, is titled "Some Strangeness in the Proportion."

19. Marjorie Nicolson, *Science and Imagination* (New York: Cornell University Press, 1962), 2.

20. Philip and Emily Morrison, "Heinrich Hertz," *Scientific American* (June 1957): 100.

The four equations are:

$$\text{div } D = \rho_f$$
$$\text{div } B = 0$$
$$\text{curl } E = -\frac{\partial B}{\partial t}$$
$$\text{curl } H = j_r + \frac{\partial D}{\partial t}$$

21. Alfred O'Rahilly, *Electromagnetic Theory* (New York: Dover Publications, 1965), 1:79. Certainly O'Rahilly's view was that "Never did a great physicist throw out such a mass of incoherent ideas, calmly pursuing his course with instinctive genius amid a welter of discrepant theories."

22. The sense of increased simplicity is reinforced by the thought that action at a distance itself could not be properly explained without supposing the existence of some medium! Newton, himself, in connection with his gravitational theory wished to anchor his force on some cause that was philosophically acceptable.

23. SU(2) is one of the many groups of symmetry operations prevalent in physics and stands for special unitary group in two dimensions. Now there is a (geometric) group of rotations that can operate on an equilateral triangle, say, so that the symmetry of the figure is retained after some of these rotations are performed. In a (mathematically) similar way the (algebraic) group SU(2) "operates" on strong, dynamical interactions between elementary particles so that total isotopic spin, related to charge, is conserved.

24. See diagram at top of page 237

25. Heisenberg, *The Physicist's Conception of Nature*, 273.

26. P,C,T stand for particular symmetry operations: P for parity—that is, mirror reflection or inversion of coordinates; C for charge conjugation or particle-antiparticle interchange; and T for time reversal.

27. Yang, *The Physicist's Conception of Nature*, 452, n. 2.

28. Even so we note that quantum theory, in the most general description, provides evidence for the discreteness of events, which is a nonaesthetic ele-

Note 24:

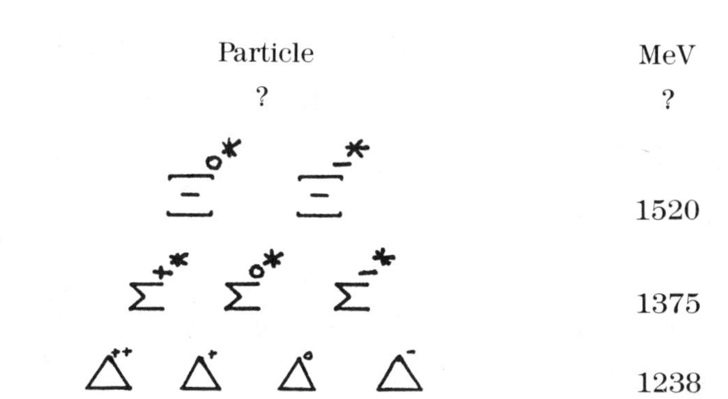

Particle	MeV
?	?
Ξ^{0*} Ξ^{-*}	1520
Σ^{+*} Σ^{0*} Σ^{-*}	1375
Δ^{++} Δ^{+} Δ^{0} Δ^{-}	1238

The obvious gap at the top of the symmetrical arrangement led to the equally obvious prediction of a particle of negative charge, and mass 1675 MeV. The Ω^{-} of mass 1675 was in fact detected.

ment, whereas at the same level relativity provides evidence for unity that does impinge on the aesthetic categories.

29. Albert Einstein, *Ideas and Opinions* (New York: Dell Publishing Co., 1973). Einstein, in *Ideas and Opinions,* wrote the following:

"It is the grand object of all theory to make these irreducible elements as *simple* and as few in number as possible."

"Our experience hitherto justifies us in believing that nature is the realisation of the *simplest* conceivable mathematical ideas."

"If I assume a Riemannian metric in it [the physical world] and ask what are the *simplest* laws which such a metric can satisfy, I arrive at the relativistic theory of gravitation in empty space."

"In the limited number of the mathematically existent *simple* field types, and the *simple* equations possible between them, lies the theorist's hope of grasping the real in all its depth."

For Einstein simplicity was concerned with harmony, beauty, and symmetry. We note that symmetry may be mathematical or physical, geometrical or conceptual, and that Einstein quite often referred to physical asymmetries. He did not think that he was doing anything unusual or invalid, for he was aware that the use of aesthetic elements had played an important role in the selection and evaluation of theories since time immemorial.

30. G. Holton, "Einstein's Scientific Program: The Formative Years," in *Some Strangeness in the Proportion,* ed. H. Woolf (Boston: Addison Wesley, 1980), 50.

31. Ibid., 59.

32. Ibid., 58.

33. S. Chandrasekhar, "Beauty and the Quest for Beauty in Science," *Physics Today* (July 1979): 25. The simplicity and the surprise are linked by Heisenberg himself who repeated a dialogue with Einstein: "you [Einstein] must have felt this too: the almost frightening simplicity and wholeness of the relationships which nature suddenly spreads out before us and for which none of us was in the least prepared."

34. Dyson, *Some Strangeness in the Proportion,* 499. Dyson, in a panel

discussion, suggested that "the equality of area and entropy [in the physics of black holes] is . . . as surprising . . . as Einstein's identification of mass with energy."

35. J. L. Synge, *Relativity: The General Theory* (Amsterdam: North Holland, 1960), preface. See also S. Goldberg, "In Defense of Ether: The British Response to Einstein's Special Theory of Relativity, 1905–1911," in *Historical Studies in the Physical Sciences* (Philadelphia: University of Pennsylvania Press, 1970), 2:99, 121. N. R. Campbell, for example, is quoted as saying that the average British physicist was "still ignorant of Einstein's work." Physicists of great ability "confess cheerfully to a complete inability to understand relativity."

36. Chrandrasekhar, "The Quest for Beauty," 26.

37. R. Carnap, *Philosophical Foundations of Physics* (New York: Basic Books, 1966), 163–64.

38. R. Angel, *Relativity* (Oxford: Pergamon Press, 1980), 197.

39. G. Holton, "What Precisely is 'Thinking'? Einstein's answer," in *Einstein,* ed. A. French (London: Heinemann, 1979), 154.

40. Although his special theory was not a constructive theory as Einstein himself noted, we can still say that the core of the theory, the Lorentz transformation equations, were derived from the two postulates together with a definition for the meaning of (distant) simultaneity.

41. General Relativity certainly owes some of its acclaim both to the fame of Einstein and to the priority given to his solution, but the intellectual feat in arriving at that particular solution, Einstein's individual style of elegance, fueled that acclaim also.

42. Mehra, *The Physicist's Conception of Nature,* 106.

43. Ibid., 108.

44. French, *Einstein,* 111.

45. T. Hutchinson, ed., *The Poetical Works of Wordsworth* (Oxford: Clarendon Press, 1936), 509. We can see him as Newton was seen by Wordsworth as "the marble index of a mind forever voyaging through strange seas of thought alone."

46. W. Watson, *Understanding Physics Today* (Cambridge: Cambridge University Press, 1963), 2.

47. G. H. Hardy, *A Mathematician's Apology* (Cambridge: Cambridge University Press, 1967), 90, 113. Our emphasis on physical theories should not obscure the fact that mathematics plays a major part in providing aesthetic ballast. G. H. Hardy, a pure mathematician of repute who wrote a beautiful and exciting little book "A Mathematician's Apology" makes no apology for insisting on the aesthetic in mathematics. At the same time he hints at the connotative as well as, more specifically, at the element of surprise. Thus he believes that the "beauty of a mathematical theorem *depends* a great deal on its seriousness, as even in poetry the beauty of a line may depend to some extent on the significance of the ideas which it contains." He then goes on to say that in both Pythagoras's and Euclid's "theorems . . . there is a very high degree of *unexpectedness,* combined with *inevitability* and *economy.* The arguments take so odd and surprising a form; the weapons used seem so childlishly simple when compared with the far-reaching results." The aesthetic in science goes beyond the mathematics as we can appreciate from Galileo's Dialogues, but without mathematics the aesthetic richness of science would be severely diminished.

48. A. Einstein and M. Besso, *Correspondence 1903–1955,* ed. P. Speziali (Paris: Hermann, 1972).

49. Woolf, *Some Strangeness in the Proportion,* 65.

50. Hardy, *Mathematician's Apology,* 119.

51. R. Dicke, "Gravitational Theory and Observation," *Physics Today* (January 1967): 55.

52. P. A. M. Dirac "The Evolution of the Physicist's Picture of Nature," *Scientific American* (May 1963): 47.

53. G. Gamow, "Does Gravity Change with Time," *Proceedings of the National Academy of Science* 57 (February 1967): 192.

Chapter 5. Science as Method

1. Preserved Smith, *The Enlightenment, 1687–1776* (New York: Collier Books, 1962), 309.

2. Indeed even if one allowed for two cultures they would not be Snow's, for the difference between them is marginal. A much greater divide exists between those who know and those who do not.

3. E. McMullin, ed., *Galileo, Man of Science* (New York: Basic Books, 1967), 7.

4. Winnifred Wisan, "Galileo and the Process of Scientific Creation," *Isis* 75 (June 1984): 269.

5. E. McMullin, "The Conception of Science in Galileo's Work," in *New Perspectives on Galileo,* ed. R. Butts and J. Pitts (Boston: Reidel Publishing Company, 1978), 211.

6. F. Bacon, *The New Organon,* ed. F. Anderson (New York: Bobbs-Merrill Co. Inc., 1960), 3.

7. Ibid., 33.

8. Ibid.

9. Ibid.

10. Ibid., 36.

11. Ibid., 39.

12. G. W. Sypher, "Similarity between the Scientific and the Historical Revolutions at the end of the Renaissance," *Journal of the History of Ideas* 26 (July–September 1965): 153.

13. Smith, *The Enlightenment,* 309.

14. Cassirer, *The Philosophy of the Enlightenment,* 53–54.

15. Voltaire, *Voltaire's Correspondence,* ed. T. Besterman (Geneva 1954), 4:48–49.

16. Bernard de Fontenelle, *Textes Choisis,* ed. M. Roelens (Paris: Editions Sociales, 1966), 280.

17. Amédée Fayol, *Fontenelle* (Paris: Nouvelle Editions Debresse, 1961), 65.

18. Cassirer, *The Philosophy of the Enlightenment,* 46, 80.

19. ". . . the distinctive feature of the period is assuredly the importance which the public accorded, from then on, to scientific subjects, and especially to generalisation." (My translation)

20. Fontenelle, *Textes Choisis,* 277.

21. Ibid., 273.

22. Ibid., 257.

23. Ibid., 258. "A man of that century [eighteenth] possesses ten times the learning of a man of Augustus' time, but he has had ten times the means for becoming learned." (My translation)

24. Smith, *The Enlightenment,* 242

25. J. P. d'Alembert, "Eléments de Philosophie," in *Mélanges de Littérature, d'Histoire, et de Philosophie* (Amsterdam, 1759), 4:3–6.

26. Preserved Smith, *A History of Modern Culture* (London: Routledge and Sons, 1930), 1:17.

27. P. B. Shelley, *A Defence of Poetry* (New York: Bobbs-Merrill, 1904), 74.

28. Ibid., 76.

29. Hutchinson, *The Poetical Works of Wordsworth,* 738.

30. Thomas Carlyle, *Sartor Resartus* (London: Chapman and Hall, 1874), 46. We note how "method" figures prominently whether a positive or a negative view of science is being expressed.

31. Matthew Arnold, *Culture and Anarchy,* ed. J. Dover Wilson (Cambridge: Cambridge University Press, 1969), 49.

32. Matthew Arnold, "Literature and Science," in *Four Essays on Life and Letters,* ed. E. Brown (New York: F. S. Crofts, 1947), 94.

33. Editorial, *Nature* (September 1870): 377.

34. Thomas Huxley, *Science and Education Essays* (New York: Appleton and Company, 1900), 141.

35. Ibid.

36. Arnold, *Culture and Anarchy,* xi.

37. Arnold, *Four Essays on Life and Letters,* 104.

38. C. P. Snow, *The Two Cultures and the Scientific Revolution* (Cambridge: Cambridge University Press, 1959)

39. C. P. Snow, "The Case of Leavis and the Serious Case," *Times Literary Supplement,* 9 July 1970, 738–40.

40. Ibid.

41. F. R. Leavis, "Literarism versus 'Scientism': The Misconception and the Menace," *Times Literary Supplement,* 30 April 1970, 441–44.

42. F. R. Leavis and M. Yudkin, *Two Cultures?* (London: Chatto and Windus, 1962), 27.

43. H. Levin, "Science and Man's Nature," in *Science and Culture,* ed. G. Holton (Boston: Beacon Press, 1967), 1ff.

44. Sir Peter Medawar, "Anglo-Saxon Attitudes," *Encounter* 25 (February 1965): 56–57.

45. The telescope functioned as more than a mere tool, for "the poetic and religious imagination of the seventeenth century was . . . changed by . . . the "new astronomy." New figures of speech appear, new themes for literature are found, new attitudes towards life are experienced." This is the sense in which the telescope is of cultural rather than mere instrumental significance.

46. René Dubos, *Science and Culture,* 253.

47. Aldous Huxley, *Literature and Science* (London: Chatto and Windus, 1963), 97.

48. Thomas Huxley, *Science and Education Essays,* 45–46.

49. J. H. Plumb, ed., *Crisis in the Humanities* (London: Pelican, 1964), 26–29.

50. Of course, he subsequently withdrew this requirement.

Chapter 6. Time and Reality in Eliot and Einstein

1. Milič Čapek, "The Myth of Frozen Passage: The Status of Becoming in the Physical World," *Boston Studies in the Philosophy of Science* (New York: Humanities Press, 1965), 2:441.

2. Sir Arthur Eddington, *The Nature of the Physical World* (London and Glasgow: Collins Clear-Type Press, 1928), 51.

3. Most people readily grasp that there is no real distinction between the three spatial dimensions or directions represented by x, y, and z. In relativity an interpretation exists that allows for no real distinction between x, y, and z on the one hand and t on the other: *t* is to be regarded simply as another, a fourth dimension. Even though the mathematics might seem to imply this equivalence of time to a spatial dimension it can readily be shown that there is not an "indissoluble mixture" of x,y,z and t: time *t* can be retrieved.

4. J. W. Dunne, *An Experiment with Time* (London: Faber and Faber, 1936): 134–40

5. Čapek, *Boston Studies,* 454–55. Čapek argues that compared with a three-dimensional (spatial) separation of past and future in the Newtonian universe, there is a four-dimensional (space-time) wedge (or Elsewhere) that serves this function in the relativistic universe. And, even at the pictorial level it is seen that "the past is separated from the future *even more effectively* than in the classical scheme!"

6. *Complete Poems and Plays of T. S. Eliot* (London: Faber and Faber, 1969), 590–91.

7. Ibid, 160.

8. T. S. Eliot, *Sacred Wood* (London: Methuen, 1960), 117.

9. Eliot, "Ash Wednesday," in *Complete Poems and Plays,* 89. Immediately following quotations are from the same collection.

10. St. Augustine, "Some Questions about Time," in *The Philosophy of Time,* ed. R. Gale (London: Macmillan, 1968), 48.

11. Ibid., 44.

12. Ibid., 40.

13. Henri Bergson, *Time and Free Will* (London: Allen and Unwin, 1950), 98.

14. Henri Bergson, *Matter and Memory* (London: Allen and Unwin, 1913), 178.

15. Henri Bergson, *Creative Evolution* (London: Allen and Unwin, 1913), 320–47.

16. T. S. Eliot, "Mr. Middleton Murry's Synthesis," *The Criterion,* 6 (1927): 346.

17. Čapek, *Boston Studies,* 447

18. T. S. Eliot, *Knowledge and Experience in the Philosophy of F. H. Bradley* (London: Faber and Faber, 1964), 111, 165.

19. Ibid., 164.

20. Ibid.

21. Ibid., 165.

22. Ibid., 158.

23. Ibid.

24. Ibid., 31.

25. Dunne, *Experiment with Time,* 106, 144–47.

26. Eliot, *Knowledge and Experience,* 10.

27. Richard Wollheim, "Eliot and F. H. Bradley," in *Eliot in Perspective: A Symposium,* ed. Graham Martin, (London: Macmillan and Company Ltd., 1970), 185.

28. Arapura G. George, *T. S. Eliot: His Mind and Art* (Inida: Asia Publishing House, 1962), 64.

29. J. T. Fraser, "The Concept of Time in Western Thought," *Main Currents in Modern Thought* 28 (March–April 1972): 115.

30. Steven Foster, "Relativity and *The Waste Land*: A Postulate," in *A Collec-*

tion of Critical Essays on "The Waste Land," ed. Jay Martin (Englewood Cliffs, N. J.: Prentice Hall, 1968), 17.

Chapter 7. The Categorical Questions in Science

1. Imre Lakatos, "History of Science and its Rational Reconstructions," *Boston Studies* (Reidel, Holland) 8 (1970): 91.

2. La Popelinière, *Histoire des histoires* (Paris, 1559), 230–31.

3. Bacon, *The New Organon,* 7–8.

4. A. C. Crombie, *Augustine to Galileo* (London: Mercury Books, 1961) 2:121.

5. David Bohm, "On the Relationship between Methodology in Scientific Research and the Content of Scientific Knowledge," *British Journal for the Philosophy of Science* 12 (August 1961): 103.

6. Sylvain Bromberger, "Why-Questions," in *Mind and Cosmos,* ed. K. Colodny (Pittsburgh: University of Pittsburgh Press, 1966), 103.

7. E. J. Dijkterhuis, *The Mechanisation of the World Picture* (Oxford: Clarendon Press, 1961), 338.

8. Crombie, *Augustine to Galileo,* 43.

9. E. Burtt, *The Metaphysical Foundations of Modern Science* (London: Routledge and Kegan Paul, 1932), 81.

10. Butts and Pitts, *New Perspectives on Galileo,* xiii.

11. Ibid., 189–89.

12. Crombie, *Augustine to Galileo,* 163.

13. E. Heller, *The Disinherited Mind* (Harmondsworth: Penguin Books, 1961), 14.

14. Ibid., 164.

15. Ibid.

16. Ibid.

17. Carl G. Hempel, *Philosophy of Natural Science* (Englewood Cliffs, N.J.: Prentice Hall, 1966), chap. 5.

18. Aristotle, *Posterior Analytics* 78a22.

19. This ignores the arguments as to whether this was a well-chosen example or even whether Aristotle's fundamental point had any validity.

20. Aristotle, *Posterior Analytics* 79a23.

21. Aristotle, *Physics* 194b20

22. Aristotle, *Metaphysics* 981a30.

23. Thomas S. Kuhn, *The Structure of Scientific Revolutions* (Chicago: University of Chicago Press, 1962), 52.

24. Indeed Popper states quite categorically that "In seeking pure knowledge our aim is, quite simply, to understand, to answer how-questions and why-questions." What-questions, from this perspective, would be asked to gain information rather than explanation.

25. McMullin, *New Perspectives on Galileo,* 216–17. McMullin asks the question in connection with Aristotle's separation of *scientia propter quid* and *scientia quia* whether *what* and *why* can really be separated. He finds that there is a disturbing circularity as one tries logically to move from any one to the other. We maintain that within normal science the separation can be made.

Within revolutionary science, however, *what* and *why* have some methodological link and the fact that we shall consider both questions to be of importance in that phase of science can be taken to mean that we acknowledge this link. However, we note that we already have considered the nature of this link in the section on Aristotle and Why questions where we saw that the questions were related but separate.

26. W. Baade and F. Zwicky, "Cosmic Rays from Super Novae," *Proceedings of the National Academy of Sciences* 20 (May 1934): 259ff.

27. J. R. Oppenheimer and H. Snyder, "On Continued Gravitational Contraction," *The Physical Review* 56 (September 1939): 455.

28. Anthony Hewish et al., "Observation of a Rapidly Pulsating Radio Source," *Nature* 217 (February 1967): 709.

29. S. Jocelyn Bell "Little Green Men, White Dwarfs, or What?" *Sky and Telescope* (March 1978): 218. Jocelyn Bell emphasizes the process of identification. On the basis of her normal skills she could say that "Fortunately, scintillation and interference usually look different . . . and one soon learns to distinguish between them." Her "scruff" did not look like either and so she had to look elsewhere for its origin.

30. Hempel, *Philosophy of Natural Science,* 54. It might be argued that it could have been otherwise; it could have been a "QUAKAR," an unknown, unpredicted "object." In Kuhn's analysis we might then be faced with an anomaly leading to revolutionary science; but this does not deny that scientists operate within a "normal" context until they are forced to do otherwise. Just such a pair of circumstance arose with respect to the normal gravitational theory of Newton. In the first case normal science was applied and Uranus was thereby *detected.* In the second case the known normal techniques did not succeed. The resulting anomaly was solved by Einstein's revolutionary theory of gravitation.

31. W. Berkson, *Fields of Force* (London: Routledge and Kegan Paul, 1974), 282.

32. Ibid., 267–68.

33. A. Einstein et al., *The Principle of Relativity* (New York: Dover Publications, 1952), 12–13.

34. H. A. Lorentz, *Collected Papers* (The Hague: Nijhoff, 1937), 4:221–22.

35. H. A. Lorentz, *The Theory of Electrons* (New York: G. E. Stechert, 1909), 230.

36. G. Holton, *Thematic Origins of Scientific Thought* (Cambridge, Mass.: Harvard University Press, 1973), 242. We can see that Kepler had a similar aim concerning "the number, distances and motions of the heavenly bodies, as to which I searched zealously for reasons why they were as they were and not otherwise." His program did not fully succeed for it was Newton whose dynamics eventually gave the answers to these why-questions.

37. Ibid., 327

38. Ibid.

39. R. Shankland, "Conversations with Albert Einstein," *American Journal of Physics* 31 (January 1963): 47.

40. P. Schilpp, ed., *Albert Einstein: Philosopher-Scientist* (Evanston, Illinois: The Library of Living Philosophers, 1949), 53.

41. Ibid.

42. In connection, for example, with the asymmetry in electrodynamics with which Einstein opened his paper on special relativity he was not concerned

with the physics; he was not going to tinker with it. It was the larger, more fundamental question of asymmetry with which he was concerned, and for the removal of this defect, only conceptual innovations rather than problem solving was the answer.

43. Holton, *Thematic Origins,* 356
44. Oresme, *A Source Book in Medieval Science,* 243–53 passim, esp. n. 24.
45. Dijkterhuis, *Mechanisation of the World Picture,* 338.
46. Ibid.

Chapter 8. Pulsar Research as Normal Science

1. Baade and Zwicky, "Cosmic Rays," 259.
2. Félix Cernuschi, "Super-Novae and the Neutron-Core Stars," *The Physical Review* 56 (July 1939): 120.
3. The statistics or the counting of particles to determine their energy, for example, depends on whether their intrinsic spin or angular momentum is described by integers or half-integers. They then obey Bose-Einstein or Fermi-Dirac statistics. Electrons and nucleons fall into the latter category, and such fermions will occupy, below a certain temperature, all the available energy states without residue, and are then described as degenerate. In a white dwarf star, for example, electrons can be treated like a gas of fermions or as a Fermi gas. In the same way that molecules exert pressure, and radiation exerts pressure, such a (degenerate) Fermi gas would also exert a pressure, *entirely related to its degeneracy.* Neutrons (in a neutron star) also exert such a pressure.
4. S. Chandrasekhar, *An Introduction to the Study of Stellar Structure* (New York: Dover Publications, 1957), 423.
5. F. Zwicky, "On the Theory and Observation of Highly Collapsed Stars," *The Physical Review* (April 1939): 726.
6. L. Landau, "Origin of Stellar Energy," *Nature* 141 (February 1938): 333.
7. G. Gamow and E. Teller, "Rate of Selective Thermonuclear Reactions," *The Physical Review* 53 (April 1938): 608.
8. J. R. Oppenheimer and R. Serber "On the Stability of Stellar Neutron Cores," *The Physical Review* 54 (October 1938): 540.
9. J. R. Oppenheimer and G. Volkoff, "On Massive Neutron Cores," *The Physical Review* 55 (February 1939): 374.
10. Oppenheimer and Snyder, "Gravitational Contraction," 455.
11. G. Gamow, "Nuclear Energy Sources and Stellar Evolution," *The Physical Review* 53 (April 1938): 595.
12. Hans Bethe, "Energy Production in Stars," *The Physical Review* 55 (March 1939: 434.
13. E. Margaret Burbidge et al., "Synthesis of Elements in Stars," *Review of Modern Physics* 29 (October 1957): 547.
14. F. Hoyle and W. Fowler, "Nucleosynthesis in Supernovas," *Astrophysical Journal* 132 (November 1960): 565. However, the simple division into two types of supernovae, I and II, as explained by Hoyle and Fowler, began to be doubted after the actual observations and measurements on the supernova explosion of 1987.
15. F. Hoyle, et al., "On Relativistic Astrophysics," *Astrophysical Journal* 139 (April 1964): 909.

16. F. Hoyle, J. Narlikar, and J. Wheeler, "Electromagnetic Waves from Very Dense Stars," *Nature* 203 (August 1964): 914.

17. D. Meltzer and K. Thorne, "Normal Modes of Radial Pulsation of Stars at the End Point of Thermonuclear Evolution," *Astrophysical Journal* 145 (August 1966): 514.

18. J. Wheeler, *Annual Review of Astronomy and Astrophysics* 4 (Palo Alto, Calif.: Annual Review Inc., 1966).

19. F. Pacini, "Energy Emission from a Neutron Star," *Nature* 216 (November 1967): 567.

20. Hewish et al., "Radio Source."

21. F. Pacini and E. Salpeter, "Some Models for Pulsed Radio Sources," *Nature* 218 (May 1968): 733.

22. T. Gold, "Rotating Neutron Stars as the Origin of the Pulsating Radio Sources," *Nature* 218 (May 1968): 731.

23. M. Large, A. Vaughn, and B. Mills, "A Pulsar Supernova Association?" *Nature* 220 (October 1968): 340.

24. Kuhn, *Structure of Scientific Revolutions,* 180.

25. D. O. Edge, "Lessons from the History of Radio Astronomy," *Proceedings of the International Conference on Using History of Physics in Innovatory Physics Education,* eds. F. Bevilacqua and P. Kennedy (1983): 145.

26. F. G. Smith, *Pulsars* (Cambridge: Camabridge University Press, 1977), xi.

27. A. G. Cameron, *Physics of Dense Matter,* ed. C. Hansen and L. Volsky (Dordrecht: Reidel, 1974), 321.

28. Steven Weinberg, *Gravitation and Cosmology* (New York: Wiley and Sons Inc., 1972), 463–64.

29. T. S. Kuhn, *Criticism and the Growth of Knowledge,* ed. T. Lakatos and A. Musgrave (Cambridge: Cambridge University Press), 234.

30. Edge, *History of Physics in Physics Education,* 148.

31. R. C. Tolman, "Static Solutions of Einstein's Field Equations for Spheres of Fluid," *The Physical Review* 55 (February 1939): 364.

32. Oppenheimer and Volkoff, "Neutron Cores."

33. Kip Thorne and James Ipser, "White-Dwarf and Neutron-Star Interpretations of Pulsating Radio Sources," *Astrophysical Journal* 152 (April 1968): 71. See also B. Bertotti, A. Cavaliere and F. Pacini, "Rotating Neutron Stars and Pulsar Emission," *Nature* 221 (February 1969): 624.

34. B. Eastlund "Low Mode Coherent Synchrotron Radiation and Pulsar Phenomena," *Nature* 225 (January 1970): 430.

35. Cohen and Börner, *Physics of Dense Matter,* 238.

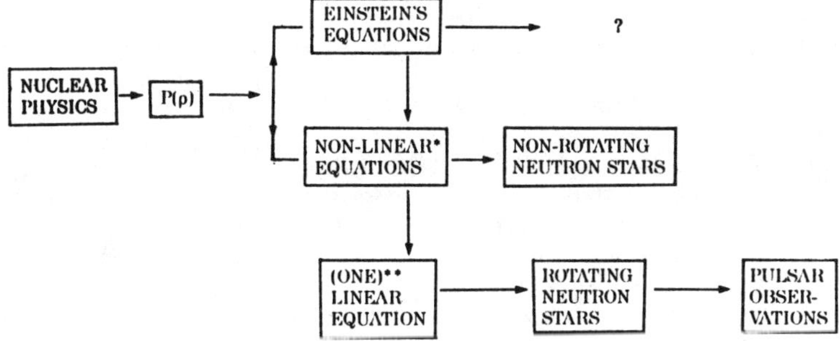

*Due to Tolman, Oppenheimer and Volkoff
**Due to Brill and Cohen

36. J. S. Hey, *The Radio Universe* (Oxford: Pergamon Press 1971), 1.

37. Ibid.

38. That 'Radio Astronomers' widened into 'High-Energy Astrophysicists' does not again defeat the thesis of continuity, for the two named groups have a great deal of cognitive overlap.

39. Weinberg, *Gravitation and Cosmology,* 317

40. Kuhn, *Structure of Scientific Revolutions,* 36.

41. Ibid., 26.

42. H.-Y. Chui and F. Occhionero, "Unified Models for Pulsars," *Nature* 223 (September 1969): 1113. See also P. Sturrock, "Pulsar Radiation Mechanisms," *Nature* 227 (August 1970): 465 and Eastlund, "Radiation and Pulsar Phenomena."

43. Kuhn, *The Structure of Scientific Revolutions,* 10.

44. Ibid., 35.

45. Ibid.

46. Kuhn, *Criticism and Growth of Knowledge,* 4.

47. News and Views, *Nature* 221 (February 1969): 710.

48. Kuhn, *Criticism and Growth of Knowledge,* 6

49. The disagreement between Kuhn and Popper may be more sharply and perspicuously put by noticing that whereas Kuhn's position may be regarded as *descriptive* and therefore unchallengeable if true, Popper regards the same position as *normative,* therefore challengeable, and challenged on the basis of his own perceptions and prejudices. It is not really a disagreement but an argument at cross purposes.

50. A. Pickering, "The Hunting of the Quark," *Isis* 72 (June 1981): 216.

51. Ibid.

52. Ibid.

53. M. Masterman, "The Nature of a Paradigm," in *Criticism and Growth of Knowledge,* 60.

54. Edge, *Using History of Physics in Physics Education,* 156

Chapter 9. From *Scientia* to Science

1. Michel Foucault, "The Human Sciences," *The Structuralists from Marx to Levi-Strauss,* ed. R. and F. De George (New York: Doubleday, 1972), 267–68. Another way to look at this change would be to accept Michel Foucault's formulation that human sciences are "part of the modern *episteme* in the same way as medicine or chemistry . . . or again, in the same way as grammar and natural history were part of the Classical *episteme,*" or even as "natural magic . . . belong[ed] to the Western episteme" before the seventeenth century. Foucault is saying that the broad epistemological network has always defined what was important, although the details keep changing.

2. Sylla, *Boston Studies* 26 (1975): 349. Sylla discusses the "commonplace of the history of science that the rise of modern science involved the breaking off of . . . specific disciplines . . . from philosophy in general. This is supposed by some to have occurred in the seventeenth century and later and by others to have had its origins in the Middle Ages." We have straddled both positions insofar as conceptions of what is science is of relevance. That science did not enjoy autonomous status until after the seventeenth century is what we would

argue, however mild the support Sylla gives to this proposition.

3. Francis Bacon, *Novum Organum* (London: William Pickering, 1850), 3–10 passim, 41.

4. Bacon, *The New Organon,* 93

5. Ibid., 43, 98. It is well known by now that Aristotle's program was not entirely an armchair exercise, but he did advocate observations as opposed to Bacon's more fruitful experiments. Furthermore, Bacon's emphasis on this element was much greater.

6. Ibid., 42. See also *Works,* 1:244.

7. Francis Bacon, *The Advancement of Learning,* ed. G. W. Kitchin (London: J. M. Dent and Sons Ltd., reprint 1954), 91–93.

8. Ibid., 100–101.

9. Ibid., 101.

10. Ibid., 90.

11. Ibid., 99.

12. Bacon, *The New Organon,* 152.

13. Ibid., 68.

14. Ibid., 115. See also *Advancement of Learning,* 33. There Bacon clearly associates "arts" with theology, logic, and mathematics and "science" with alchemy (Chemistry) and magnetism. "For these [the former] were the arts which had a kind of primogeniture with [Aristotle, Plato, and Proclus] severally"; and the others were "some sciences" through which "the alchymists [have] made a philosophy out of a few experiments of the furance; and Gilbertus . . . hath made a philosophy out of the observations of a lodestone."

15. Bacon, *The Advancement of Learning,* 90.

16. Cotter, *ABC of Scholastic Philosophy,* 1–3.

17. I. Bernard Cohen, *The Newtonian Revolution* (Cambridge: Cambridge University Press, 1980), 190–91. Cohen argues similarly about Newton's use both of the outdated, indeed the incorrect, centrifugal (instead of centripetal) force in his *Principia,* as well as of the Aristotelian *vis insita.* Cohen believes that it is an example of the "persistence of . . . phrases that remain long after the concepts for which they . . . stood have been rejected. They stand as signs . . . of continuity within change, a terminological invariance in a transformation." Indeed Cohen argues, from the internal evidence, that Newton knew that there was no *vis insita,* that there was no internal *force* that was the cause of a body's uniform motion.

18. More precisely, Newton engaged in these two kinds of sciences. In the *Principia* we have the full flowering of the classical sciences; it is in the *Opticks* that Newton effects the bridge between the two sciences.

19. Editorial, *New Scientist,* 14 July 1960, 87. Indeed, Bacon is known to have advocated a research establishment in his *New Atlantis* in 1627, called by him Solomon's House. One might say that Bacon was very prescient about this particular development of science. More plausibly, it was the force of Bacon's imaginative example that impelled scholars to develop science in this particular way.

20. Other institutions like universities showed changes in their perceptions of science, sometimes reflected in their courses. Thus, in France professors began to accept the accuracy of astronomer's observations and integrate them in their courses. Brockliss believes that the change in the epistemology of science that took place in the seventeenth century lay in the willingness of these professionals to accept the conclusions of contemporary experimental

philosophers.

21. Bacon, *Advancement of Learning*, 90.

22. Bernard de Fontenelle, "Éloge de M. Varignon," in *Oeuvres de Fontenelle* (Paris: Libraires Associés, 1766), 6: 172.

23. This must be contrasted with the situation in 1601 when "physics" was taught as a section of a four-part course, always following logic and metaphysics. For logic as the tool for a knowledge of physics and metaphysics provided the fundamental principles. There was clearly an Aristotelian framework.

24. L. W. Brockliss, "Aristotle, Descartes and the New Science", *Annals of Science* 38 (1981): 33.

25. T. Martin, *The Royal Institution* (London: Oliver Burridge and Co., 1961), 14.

26. M. Kerker, "Science and the Steam Engine" in *The Development of Western Technology*, ed. T. Hughes (New York: Macmillan, 1964), 70.

27. Ibid., 72.

28. Immanuel Kant, *Metaphysical Foundations of Natural Science* (New York: Bobbs-Merrill Co., 1970), 4.

29. Ibid.

30. "Dr. Black's Theory of Latent Heat," *The Edinburgh Review* 3 (October 1803): 4.

31. "An enquiry concerning the Nature of Heat, and the mode of its Conservation," *The Edinburgh Review* 4 (July 1804): 399.

32. The fact that modern scientists frame hypotheses does not diminish the central force of this critique as it seems merely to be taking the inductivist side of the perennial debate about the growth of science.

33. *Edinburgh Review*, 1804.

34. Thomas Thompson, *History of the Royal Society* (London, 1812), 311.

35. W. Whewell, *Philosophy of the Inductive Sciences* (London: J. Parker, 1840), 1: introduction.

36. W. Whewell, *History of the Inductive Sciences* (London: J. Parker, 1857), 1:5.

37. There are still two views about Galileo's status as a (modern) experimenter, especially in connection with his "inclined planes." Thus was he really a "rationalist" who had arrived at his kinematic equations and been totally convinced of their truth, the "inclined planes" serving merely as an adjunct of his laws? Or did he believe that it was of vital importance to test these laws so that he saw the inclined planes as an important and integral part of his scientific alternative that the internal evidence seems to support and that, therefore, diminishes his role in the eyes of some scholars?

38. T. S. Kuhn, *The Essential Tension* (Chicago: Chicago University Press, 1977), 31ff.

39. C. Singer, *A Short History of Scientific Ideas* (Oxford: Oxford University Press, 1962), 134.

40. Kuhn, *The Essential Tension*, 51–52. See also M. Crosland and C. Smith, "The Transmission of Physics from France to Britain," in *Historical Studies in the Physical Sciences* (Baltimore and London: John Hopkins University Press, 1978), 9:1.

41. In the *Oxford English Dictionary*, the earliest entry in which *Science* is "contradistinguished from *art*" is 1678: "Though we may justly account Dyalling originally a Science, yet . . . it is now become . . . no more difficult than an Art."

In the same source, the earliest entry in which the modern sense of the word can be discerned is 1725: "The word Science is usually applied to a whole body of . . . methodical observations or propositions."

An important point to make is that even though the earlier (1678) entry looks promisingly modern, it was being used as an example of "science," only as a particular branch of knowledge. On the other hand, the second entry was exemplifying science as a "branch of study . . . concerned either with a connected body of demonstrated truths or with observed facts systematically classified and more or less colligated by being brought under general laws, and which includes trustworthy methods for the discovery of new truth within its own domain."

Thus, 1700 seems to be a reasonable date to suggest for the recognized acquisition by science of its modern connotations.

Chapter 10. Classical Duality in Modern Physics

1. W. R. Hamilton, *Mathematical Papers* (Cambridge: Cambridge University Press, 1931), 1:311–32. See also P. Fermat, *Oeuvres de Fermat* (Paris, 1891), 1:173, and P. Maupertuis *Les Oeuvres de Maupertuis,* vol. 1 (Berlin, 1753).

2. Basically the intention of the theorists was to explain how radiation energy was distributed over the wavelength spectrum. Rayleigh and Jeans found that their theory disagreed with the empirical data except for long wavelengths. Wien's theory disagreed rather less with the data but did so for short wavelengths.

3. Max Planck, *Planck's Original Papers in Quantum Physics* (London: Taylor and Francis, 1972), 35–37.

4. His work allowed others to see that certain quantities and processes could be better described by using the discrete rather than the continuous category.

5. R. H. Stuewer, *The Compton Effect: Turning Point in Physics* (New York: Science History, 1975), chap. 1. Steuewer has shown that wave-particle duality was discussed among x-ray physicists independently of the quantum principle.

6. Hamilton, *Mathematical Papers,* 315. "Those who have meditated on the beauty and utility, in theoretical mechanics, of the general method of Lagrange . . . [will see that the] general problem that I have proposed to myself in optics, is to investigate the mathematical consequences of the law of least action."

7. The concept of the photon alone certainly does not entail any wave theory. Furthermore MATRIX mechanics is an adequate description of atomic phenomena.

8. Those who find the mathematics taxing may move on to the conclusion and still glean enough of the purpose of this chapter, if rather less of its substance.

9. Thus, in a medium of uniform optical properties, Fermat's principle leads to the expectation of a straight line path, consistent with the common assertion that light travels in straight lines.

10. A. H. Compton, "A Quantum Theory of the Scattering of X-rays by Light Elements," *Physics Review* 21 (1923): 483.

11. C. J. Davisson and L. H. Germer, "The Scattering of Electrons by a Single Crystal of Nickel," *Nature* 119 (1927): 558. In fact, it was Elsasser who first saw the possibility of a wave interpretation of some early results of particle scattering achieved by Davisson. Elsasser suggested further experiments, which Davisson and Germer, as well as G. P. Thomson, carried out.

12. E. MacKinnon, "DeBroglie's Thesis: A Critical Retrospective," *American Journal of Physics* 44 (November 1976): 1047. There is a case that argues that deBroglie confused his representations.

Bibliography

Alembert, J. P. d'. *Mélanges de Littérature, d'Histoire, et de Philosophie,* vol. 4. Amsterdam, 1759.

Anderson, J. L. *Principles of Relativity Physics.* New York: Academic Press, 1967.

Angel, Roger B. *Relativity: The Theory and Its Philosophy.* Oxford: Pergamon Press, 1980.

Arber, Agnes. *The Mind and the Eye: A Study of the Biologist's Standpoint.* Cambridge: Cambridge University Press, 1964.

Aristotle. *The Basic Works of Aristotle.* Edited by Richard McKeon. New York: Random House, 1941.

——. *Posterior Analytics.* Translated by Hugh Tredennick. London: William Heinemann, 1960.

Arnold, Matthew. *Four Essays on Life and Letters.* Edited by E. Brown. New York: F. S. Crofts, 1947.

——. *Culture and Anarchy.* Edited by J. Dover Wilson. Cambridge: Cambridge University Press, 1969.

Baade, W., and Zwicky, F. "Cosmic Rays from Super-Novae." *Proceedings of the National Academy of Sciences* 20 (May 1934): 259–63.

Bacon, Francis. *The Advancement of Learning.* Edited by G. W. Kitchin. London: J. M. Dent and Sons Ltd., 1954.

——. *The New Organon and Related Writings.* Edited by F. Anderson. Translated by James Spedding, Robert Leslie Ellis, and Douglas Denon Heath. Indianapolis and New York: Bobbs-Merrill Co. Inc., 1960.

——. (Verulam, Francis Lord). *Novum Organum or True Suggestions for the Interpretation of Nature.* London: William Pickering, 1850.

——. *The Essays of Francis Bacon.* Edited by M. Scott. New York: Scribners' Sons, 1908.

——. *The Works of Francis Bacon,* 15 vols. Edited and compiled by James Spedding, Robert Leslie Ellis, and Douglas Devon Heath. Boston: Brown and Taggard, 1861.

Bacon, Roger. *The Opus Maius of Roger Bacon,* 2 vols. Translated by Robert Belle Burke. New York: Russell and Russell, 1962.

——. *Fr. Rogeri Bacon Opera.* Edited by J. S. Brewer. London: Longman, 1859.

Bell, S. Jocelyn. "Little Green Man, White Dwarfs, or What?" *Sky and Telescope* (March 1978): 218–21.

Bergson, Henri. *Creative Evolution.* London: Allen and Unwin, 1913.

——. *Matter and Memory.* London: Allen and Unwin, 1913.

——. *Time and Free Will.* London: Allen and Unwin, 1950.

Berkson, W. *Fields of Force.* London: Routledge and Kegan Paul, 1974.

Bertotti, B., A. Cavaliere, and F. Pacini. "Rotating Neutron Stars and Pulsar Emission." *Nature* 221 (February 1969): 624–25.

Bethe, Hans. "Energy Production in Stars." *Physical Review* 55 (March 1939): 434–56.

Bieler, Ludovicus, ed. *Corpus Christianorum Series Latina XCIV.* Brepols: Editores Pontificii, 1957.

Bevilacqua, F. and P. J. Kennedy, eds. *Proceedings of the International Conference on Using History of Physics in Innovatory Physics Education.* International Commission on Physics Education, 1984.

Birkhoff, G. D. *Aesthetic Measure.* Cambridge, Mass.: Harvard University Press, 1933.

Boethius. *The Consolations of Philosophy of Boethius.* Translated by H. R. James. London: G. Routledge and Sons.

Bohm, David. "On the Relationship between Methodology in Scientific Research and the Content of Scientific Knowledge." *British Journal for the Philosophy of Science* 12 (August 1961): 103ff.

Brockliss, L. W. B. "Aristotle, Descartes and the New Science: Natural Philosophy at the University of Paris, 1600–1740." *Annals of Science* 38 (1981): 33–69.

Broglie, Louis de. *Nobel Lectures: Physics 1922–1941.* Amsterdam: Elsevier Publishing Co. 1965, 239–59.

Bronowski, Jacob. *Science and Human Values.* New York: Harper and Row, 1965.

Bunge, Mario. "The Complexity of Simplicity." *The Journal of Philosophy* 59 (March 1962). See also *The Myth of Simplicity.* Englewood Cliffs, N.J.: Prentice Hall, 1963, chapters 4 and 5.

Bunge, Mario, ed. *The Critical Approach to Science and Philosophy.* New York: The Free Press of Glencoe, 1964.

Burbidge, E. Margaret, G. R. Burbidge, William A. Fowler, and F. Hoyle "Synthesis of the Elements in Stars." *Review of Modern Physics* 29 (October 1957): 547–650.

Burtt, Edwin Arthur. *The Metaphysical Foundations of Modern Science.* London: Routledge and Kegan Paul, 1932.

Butts, R. and J. Pitts, eds. *New Perspectives on Galileo.* Boston: Reidel Publishing Co., 1978.

Campbell, N. R. *Physics: The Elements.* Cambridge: Cambridge University Press, 1920.

Carlyle, Thomas. *Sartor Resartus: The Life and Opinions of Herr Teufelsdröckh.* London: Chapman and Hall, 1868.

Carnap, Rudolph. *Philosophical Foundations of Physics.* Edited by M. Gardner. New York: Basic Books, 1966.

Cassirer, Ernst. *The Problem of Knowledge, Philosophy, Science, and History since Hegel.* Translated by W. H. Woglom and C. H. Hendel. New Haven: Yale University Press, 1950.

———. *The Philosophy of the Enlightenment.* Translated by Fritz C. A. Koelin and James P. Pettegrove. Princeton: Princeton University Press, 1951.

Cernuschi, Félix. "Super-Novae and the Neutron-Core Stars." *The Physical Review* 56 (July 1939): 120.

Chandrasekhar, S. *An Introduction to the Study of Stellar Structure.* New York: Dover Publications, 1957.

———. "Beauty and the Quest for Beauty in Science." *Physics Today* (July 1979): 25–30.

Chaucer, Geoffrey. *The Complete Works of Chaucer*. Edited by Walter Skeat. London: Oxford University Press, 1912.

Chiu, H.-Y. and F. Occhionero "Unified Models for Pulsars." *Nature* 223 (13 September 1969): 1113–16.

Cohen, I. Bernard. *The Newtonian Revolution*. Cambridge: Cambridge University Press, 1980.

Colodny, K., ed. *Mind and Cosmos*. Pittsburgh: University of Pittsburgh Press, 1966.

Compton, A. H. "A Quantum Theory of the Scattering of X-rays by Light Elements." *The Physical Review* 21 (May 1923): 483–502.

Cotter, A. C. *ABC of Scholastic Philosophy*. Weston, Mass.: The Weston College Press, 1949.

Crombie, A. C. *Robert Grosseteste and the Origin of Experimental Science 1100–1700*. Oxford: Clarendon Press, 1953.

———. *Augustine to Galileo*, 2 vols. London: Mercury Books, 1961.

Crousaz, J. P. de. *Traité du Beau*. Amsterdam: François L'Honoré, 1715.

Dantzig, Tobias. *Number: The Language of Science*, 4th ed. New York: Free Press, 1967.

Davisson, C. J. and L. H. Germer. "The Scattering of Electrons by a Single Crystal of Nickel." *Nature* 119 (16 April 1927): 558–60.

Descartes, René. *The Philosophical Works of Descartes*, 2 vols. Translated by Elizabeth S. Haldene and G. R. T. Ross. Cambridge: Cambridge University Press, 1972.

Dicke, R. "Gravitational Theory and Observation." *Physics Today* (January 1967): 55–70.

Dijkterhuis, E. J. *The Mechanisation of the World Picture*. Translated by C. Dikshoorn. Oxford: Clarendon Press, 1961.

Dirac, Paul A. "The Evolution of the Physicists' Picture of Nature," *Scientific American* (May 1963): 45–53.

Dreyer, J. L. *History of the Planetary Systems from Thales to Kepler*. Cambridge: Cambridge University Press, 1906.

Dunne, J. W. *An Experiment with Time*, 4th ed. London: Faber and Faber, 1936.

Durrant, William. *The Story of Philosophy*. New York: Simon and Schuster, 1953.

Eastlund, B. "Low Mode Coherent Synchrotron Radiation and Pulsar Phenomnena." *Nature* 225 (31 January 1970): 430–34.

Easton, Stewart C. *Roger Bacon and his Search for a Universal Science*. Oxford: Blackwell, 1952.

Eddington, Sir Arthur. *The Nature of the Physical World*. London and Glasgow: Collins Clear-Type Press, 1928.

Einstein, Albert. *Sidelights on Relativity*. London: Methuen, 1922.

———. *Ideas and Opinions*. New York: Dell Publishing Co., 1973.

Einstein, Albert and Michele Besso. *Correspondence 1903–1955*. Edited by P. Speziali. Paris: Hermann, 1972.

Einstein, Albert, H. A. Lorentz, H. Minkowski, and H. Weyl. *The Principle of*

Relativity. Notes by A. Sommerfeld. Translated by W. Perrett and G. B. Jeffery. New York: Dover Publications, 1952.

Eliot, T. S. "Mr. Middleton Murry's Synthesis." *The Criterion* 6 (December 1927): 340–47.

———. *Sacred Wood.* London: Methuen, 1960.

———. *Knowledge and Experience in the Philosophy of F. H. Bradley.* London: Faber and Faber, 1964.

———. *Complete Poems and Plays of T. S. Eliot.* London: Faber and Faber, 1969.

Faraday, Michael. *Experimental Researches in Chemistry and Physics.* London: Taylor, 1859.

———. *Experimental Researches in Electricity,* 3 vols. Edited by R. Taylor and W. Francis. London, 1839–1855. Reprint, New York: Dover Publications, 1965.

Fayol, Amédée. *Fontenelle.* Paris: Nouvelles Edition Debresse, 1961.

Fermat, Pierre. *Oeuvres de Fermat,* 4 vols. Edited by P. Tannery and C. Henry. Paris, 1891–1912.

Fontenelle, Bernard de, *Oeuvres de Monsieur De Fontenelle,* 11 vols. Paris. Libraries Associés, 1766.

———. *Textes Choisis, 1683–1702.* Introduction and notes by Maurice Roelens. Paris: Editions Sociales, 1966.

Fraser, J. T. "The Concept of Time in Western Thought." *Main Currents in Modern Thought* 28 (March–April 1972): 111–24.

French, A. P., ed. *Einstein: A Centenary Volume.* London: Heinemann, 1979.

Gale, Richard M., ed. *The Philosophy of Time.* London: Macmillan & Co., 1968.

Gamow, George. "Nuclear Energy Sources and Stellar Evolution." *The Physical Review* 53 (April 1938): 595–604.

———. "Does Gravity Change with Time?" *Proceedings of the National Academy of Sciences* 57 (February 1967): 187–93.

Gamow, George, and E. Teller. "Rate of Selective Thermonuclear Reactions." *The Physical Review* 53 (April 1938): 608–9.

George, Arapura G. *T. S. Eliot: His Mind and Art.* Bombay: Asia Publishing House, 1962.

George, Richard de, and Fernande de George, eds. *The Structuralists from Marx to Levi-Strauss.* New York: Doubleday, 1972.

Gold, T. "Rotating Neutron Stars as the Origin of the Pulsating Radio Sources." *Nature* 218 (25 May 1968): 731–32.

Goldberg, Leo, and David Layzer, and John G. Phillips, eds. *Annual Review of Astronomy and Astrophysics.* 4. Palo Alto, Calif.: Annual Review Inc., 1966.

Goodman, Nelson. "Recent Developments in the Theory of Simplicity." *Philosophical and Phenomenological Research* 19 (June 1959): 429–46.

Grant, Edward, ed. *A Sourcebook in Medieval Science.* Cambridge, Mass.: Harvard University Press, 1974.

Hamilton, W. R. *The Mathematical Papers of Sir William Rowan Hamilton.* Edited by A. W. Conway and J. L. Synge. Vol. 1. *Geometrical Optics.* Cambridge: Cambridge University Press, 1931.

Hansen, C., and L. Volsky, eds. *Physics of Dense Matter.* Dordrecht: Reidel Publishing Co., 1974.

Hardy, G. H. *A Mathematician's Apology.* Cambridge: Cambridge University Press, 1967.

Heller, Erich. *The Disinherited Mind.* Harmondsworth, Middlesex: Penguin Books Ltd., 1961.

Hempel, Carl G. *Philosophy of Natural Science.* Englewood Cliffs, N.J.: Prentice Hall, 1966.

Hewish, A., S. J. Bell, J. D. H. Pilkington, P. F. Scott, and R. A. Collins. "Observation of a Rapidly Pulsating Radio Source." *Nature* 217 (February 1967): 709–13.

Hey, J. S. *The Radio Universe.* Oxford: Pergamon Press, 1971.

Holton, Gerald. *Thematic Origins of Scientific Thought.* Cambridge, Mass.: Harvard University Press, 1973.

Holton, Gerald, ed. *Science and Culture.* Boston: Beacon Press, 1967.

Hoyle, F. and W. Fowler. "Nucleosynthesis in Supernovae." *Astrophysical Journal* 132 (November 1960): 565–90.

Hoyle, F., W. Fowler, G. R. Burbidge, and E. Margaret Burbidge. "On Relativistic Astrophysics." *Astrophysics Journal* 139 (April 1964): 909–28.

Hoyle, F., J. Narlikar, and J. Wheeler. "Electromagnetic Waves from Very Dense Stars." *Nature* 203 (29 August 1964): 914–16.

Hughes, Thomas Parke, ed. *The Development of Western Technology.* New York: Macmillan and Co., 1964.

Huxley, Aldous. *Literature and Science.* London: Chatto and Windus, 1963.

Huxley, Thomas. *Science and Education Essays.* New York: Appleton and Co., 1900.

Jammer, Max. *The Conceptual Development of Quantum Mechanics.* New York: McGraw-Hill, 1966.

Kant, Immanuel. *Metaphysical Foundations of Natural Science.* Translated by James Ellington. New York: Bobbs-Merrill Co. Inc., 1970.

Kepler, Johannes. *Paralipomena ad Vitellionem.* Frankfurt, 1604.

Koestler, Arthur. *The Sleepwalkers.* New York: Grosset and Dunlop, 1963.

Komesaroff, M. "Possible Mechanism for the Pulsar Radio Emission." *Nature* 225 (14 February 1970): 612–14.

Koyré, Alexander. *Newtonian Studies.* Chicago: University of Chicago Press, 1965.

Kuhn, Thomas S. *The Structure of Scientific Revolutions.* Chicago: University of Chicago Press, 1962.

———. *The Essential Tension.* Chicago: University of Chicago Press, 1977.

Lakatos, Imre, and A. Musgrave, eds. *Criticism and the Growth of Knowledge.* Cambridge: Cambridge University Press, 1970.

Landau, L. D. "Origin of Stellar Energy." *Nature* 141 (19 February 1938): 3

Large, M. A. Vaughn, and B. Mills. "A Pulsar Supernova Association?" *Nature* 220 (26 October 1968): 340–41.

Leavis, F. R. "Literarism versus 'Scientism': The Misconception and the Menace." *Times Literary Supplement,* London (30 April 1970): 441–44.

Leavis, F. R., and M. Yudkin. *Two Cultures?* London: Chatto and Windus, 1962.

Lindberg, David C. "Alkindi's Critique of Euclid's Theory of Vision." *Isis* 62 (Winter 1971): 469–89.

Lorentz, H. A. *Collected Papers,* 9 vols. Edited by P. Zeeman and A. D. Fokker. The Hague: Nijhoff, 1934–38.

———. *The Theory of Electrons.* New York: G. E. Stechert, 1909. New York: Dover Publications, 1952.

Magie, William Francis. *A Sourcebook in Physics.* Cambridge, Mass.: Harvard University Press, 1969.

Marsak, Leonard M., ed. *The Rise of Science in Relation to Society.* New York: The Macmillan Co., 1964.

Martin, Graham, ed. *Eliot in Perspective: A Symposium.* London: Macmillan and Co. Ltd., 1970.

Martin, Jay, ed. *A Collection of Critical Essays on "The Waste Land."* Englewood Cliffs, N.J.: Prentice Hall, 1968.

Martin, T. *The Royal Institution.* London: Oliver Burridge and Co., 1961.

Maupertuis, P. L. M. de "The Agreement Between the Different Laws of Nature that Had, Until Now, Seemed Incompatible." *Mémoires de l'Académie des Sciences.* Paris, April 1774.

Mac Kinnon, Edward. "De Broglie's Thesis: A Critical Retrospective." *American Journal of Physics* 44 (November 1976): 1047–55.

McCormach, Russell, ed. *Historical Studies in the Physical Sciences,* vol. 2. Philadelphia: University of Pennsylvania Press, 1970.

McCormach, Russell, and Lewis Pyenson eds. *Historical Studies in the Physical Sciences,* vol. 9. Baltimore and London: The John Hopkins University Press, 1978.

McGarry, Daniel D., trans. and ed. *The Metalogicon of John of Salisbury: A Twelfth-Century Defense of the Verbal and Logical Arts of the Trivium.* Berkeley and Los Angeles: University of California Press, 1955.

McMullin, Ernan, ed. *Galileo, Man of Science.* New York: Basic Books, 1967.

Medawar, P. B. "Anglo-Saxon Attitudes." *Encounter* 25 (August 1965): 52–58.

Mehra, Jagdish, ed. *The Physicist's Conception of Nature.* Dordrecht: D. Reidel Publishing Co., 1973.

Meltzer, D., and Thorne, K. "Normal Modes of Radial Pulsation of Stars at the End Point of Thermonuclear Evolution." *Astrophysical Journal* 145 (August 1966): 514–43.

Molland, A. G. "Medieval Ideas of Scientific Progress." *Journal of the History of Ideas* 39 (October–December 1978): 561–77.

Morrison, Philip, and Emily Morrison. "Heinrich Hertz." *Scientific American* (June 1957): 98–106.

Murdoch, John Emery, and Edith Dudley Sylla eds. *The Cultural Context of Medieval Learning.* Boston Studies 26. Boston: Reidel Pub. Co., 1975.

Newton, Sir Isaac. *Opticks.* New York: Dover Publications, 1952.

———. *Isaac Newton's Papers and Letters on Natural Philosophy and Related Documents.* Edited by I. Bernard Cohen and Robert E. Schofield, Cambridge: Cambridge University Press, 1958.

Nicolson, Marjorie. *Science and Imagination.* New York: Cornell University Press, 1962.

Oppenheimer, J. R., and R. Serber. "On the Stability of Stellar Neutron Cores." *The Physical Review* 54 (1 October 1938): 540.

Oppenheimer, J. R., and H. Snyder. "On Continued Gravitational Contraction." *The Physical Review* 56 (1 September 1939): 455–59.

Oppenheimer, J. R., and G. Volkoff. "On Massive Neutron Cores." *The Physical Review* 55 (15 February 1939): 374–81.

O'Rahilly, Alfred. *Electromagnetic Theory: A Critical Examination of Fundamentals.* New York: Dover Publications, 1965.

Pacini, F. "Energy Emission from a Neutron Star." *Nature* 216 (11 November 1967): 567–68.

Pacini, F., and E. Salpeter. "Some Models for Pulsed Radio Sources." *Nature* 218 (25 May 1968): 733–34.

Peacock, W., ed. *English Prose,* 5 vols. London: Oxford University Press, 1921.

Pickering, Andrew. "The Hunting of the Quark." *Isis* 72 (June 1981): 216–36.

Planck, Max. *Planck's Original Papers in Quantum Physics.* Annotated by Hans Kangro. Translated by D. ter Haar and Stephen G. Brush. London: Taylor and Francis, 1972.

Plumb, J. H. *Crisis in the Humanities.* Harmondsworth, Middlesex: Penguin Books, 1964.

Poincaré, Henri. *Science and Method.* New York: Dover Publications, 1952.

Polanyi, Michael. *Personal Knowledge: Towards a Post-Critical Philosophy.* London: Routledge and Kegan Paul, 1958.

Popelinière, Lancelot de La. *Histoire des histoires . . .* Paris, 1599.

Popper, Karl R. *Conjectures and Refutations.* London: Routledge and Kegan Pual, 1969.

———. *The Logic of Scientific Discovery,* 3rd ed. Translated by Karl Popper, Julius Freed, and Lan Freed. London: Hutchinson and Co. Ltd., 1972.

———. *Objective Knowledge: An Evolutionary Approach.* Oxford: The Clarendon Press, 1972. Reprint, 1973.

Ramsay, Frank. *The Foundations of Mathematics and Other Logical Essays.* Edited by R. B. Braithwaite. London: Routledge and Kegan Paul, 1931. Reprint, 1965.

Rawlins, F. "Methodology and Aesthetic Content in Natural Science." *The Institute of Physics Bulletin* (September 1959): 205–9.

Rindler, Wolfgang. "Relativistic Cosmology." *Physics Today* (November 1967); 23–31.

Ronchi, Vasco. *The Nature of Light.* Translated by V. Barocas. London: Heinemann, 1970.

Ross, Sydney. "The Study of a Word." *Annals of Science* 18 (1962): 65–85.

Ruskin, John. *The Works of John Ruskin,* 39 vols. Edited by E. R. Cook and Alexander Weddeburn. London: George Allen, 1903–11.

Sabra, A. I. *Theories of Light from Descartes to Newton.* London: Oldbourne Book Co. Ltd., 1967.

Sambursky, S. *The Physical World of Late Antiquity.* London: Routledge and Kegan Paul, 1962.

Santayana, George. *The Sense of Beauty.* New York: Dover Publications, 1955.

Schilpp, P., ed. *Albert Einstein: Philosopher-Scientist.* Evanston, Ill.: The Library of Living Philosophers, 1949.

Schrödinger, Erwin. *Science, Theory and Man.* New York: Dover Publications, 1957.

Shankland, R. "Conversations with Albert Einstein." *American Journal of Physics* 31 (January 1963): 47–57.

Sharma, C. S. "The Role of Mathematics in Physics." *British Journal for the Philosophy of Science* 33 (September 1982): 275–86.

Shelley, Percy Bysshe. *A Defence of Poetry.* Edited by Mrs. Shelley. New York: The Bobbs-Merrill Co., 1904.

Singer, Charles. *A Short History of Scientific Ideas to 1900.* Oxford: Oxford University Press, 1962.

Smith, F. G. *Pulsars.* Cambridge: Cambridge University Press, 1977.

Smith, Preserved. *A History of Modern Culture,* 2 vols. London: Routledge and Sons, 1930.

———. *The Enlightenment, 1687–1776.* New York: Collier Books, 1962.

Snow, C. P. *The Two Cultures and the Scientific Revolution.* Cambridge: Cambridge University Press, 1959.

———. "The Case of Leavis and the Serious Case." *Times Literary Supplement,* London (9 July 1970): 738–40.

Stiefel, Tina. "The Heresy of Science: A Twelfth-Century Conceptual Revolution." *Isis* 68 (September 1977): 347–62.

Stock, Brian. *Myth and Science in the Twelfth Century: A Study of Bernard Silvester.* Princeton: Princeton University Press, 1972

Stuewer, Roger H. *The Compton Effect—Turning Point in Physics.* New York: Science History, 1975.

Sturrock, P. "Pulsar Radiation Mechanics." *Nature* 227 (1 August 1970): 465–70.

Synge, J. L. *Relativity: The General Theory.* Amsterdam: North Holland, 1960.

Sypher, G. Wylie. "Similarity between the Scientific and the Historical Revolutions at the End of the Renaissance." *Journal of the History of Ideas* 26 (July–September 1965): 353–68.

Thompson, D'Arcy. *On Growth and Form.* Abridged edition. Cambridge: Cambridge University Press, 1966.

Thomson, Thomas. *History of the Royal Society from its Institution to the End of the Eighteenth Century.* London: R. Baldwin, 1812.

Thorne, Kip, and James Ipser. "White-Dwarf and Neutron-Star Interpretations of Pulsating Radio Sources." *Astrophysical Journal* 152 (April 1968): L 71–75.

Tolman, R. C. "Static Solutions of Einstein's Field Equations for Spheres of Fluid." *The Physical Review* 55 (15 February 1939): 364–73.

Voltaire, François. *Voltaire's Correspondence,* 6 vols. Edited by T. Besterman. Genève: Institute et musée de Voltaire, 1954.

Watson, W. *Understanding Physics Today.* Cambridge: Cambridge University Press, 1963.

Weinberg, Steven. *Gravitation and Cosmology.* New York: Wiley and Sons Inc., 1972.

Weisskopf, Victor F. "Is Physics Human?" *Physics Today* (June 1976): 23–29.

Whewell, William. *Philosophy of the Inductive Sciences Founded upon Their History,* 2 vols. London: John Parker and Son, 1840.

———. *History of the Inductive Sciences from the Earliest to the Present Times,* 3 vols, 3rd ed. London: John Parker and Son, 1857.

Whyte, Lancelot Law. *Accent on Form.* New York: Harper and Row, 1954.

Whyte, Lancelot Law, ed. *Aspects of Form: A Symposium on Form in Nature and Art,* 2nd ed. London: Lund Humphries, 1968.

Wiener, Philip, and Aaron Noland, eds. *Roots of Scientific Thought.* New York: Basic Books, 1957.

Wisan, Winnifred Howell. "Galileo and the Process of Scientific Creation." *Isis* 75 (June 1984): 269–86.

Wittgenstein, Ludwig. *Lectures and Conversations on Aesthetics, Psychology and Religious Belief.* Edited by C. Barrett from notes compiled by Yorrick Smithes, Rush Rhees, and James Taylor. Berkeley: University of California Press, 1967.

Woolf, Harry, ed. *Some Strangeness in the Proportion.* Boston: Addison-Wesley, 1980.

Wordsworth, William. *The Poetical Works of Wordsworth.* Edited by T. Hutchinson. Oxford: Clarendon Press, 1904. Revised edition by Ernest de Selincourt, 1936.

Zwicky, F. "On the Theory and Observation of Highly Collapsed Stars." *The Physical Review* 55 (15 April 1939): 726–43.